System Design Approaches
to Public Services

Other Books by JOHN H. BURGESS:
The Christian Pagan
The Time Dimension in Science and Psychology

System Design Approaches to Public Services

John H. Burgess

Rutherford • Madison • Teaneck
Fairleigh Dickinson University Press
London: Associated University Presses

Associated University Presses, Inc.
Cranbury, New Jersey 08512

Associated University Presses
Magdalen House
136–148 Tooley Street
London SE1 2TT, England

Library of Congress Cataloging in Publication Data

Burgess, John H.
    System design approaches to public services.

    Bibliography: p.
    Includes index.
    1. Municipal services. 2. Social service. 3. System analysis.
I. Title.
HD4431.B87   363.5   76-737
ISBN 0-8386-1892-8

PRINTED IN THE UNITED STATES OF AMERICA

For want of a nail the shoe is lost, for want of shoe the horse is lost, for want of a horse the rider is lost. For want of a rider the battle was lost. For want of a battle the kingdom was lost. And all for want of a horseshoe nail.

—George Herbert (1593–1633)

Contents

PART I. A Systems Perspective in the Modern World

1. Introduction: The Importance of a Systems Approach

   Systems most easily identified in nature: astronomy, meteorology, biology. Man-made systems: garbage disposal, police and criminal justice, fire protection, transportation and vehicle systems, health-care systems, mental health, welfare services. Systems and human-services engineering. Analytical methods applied to human-service systems. Design and implementation.

2. Human-Service Systems and Human-Factors Methods of Analysis

   A human-services policy formulation. Systems operational requirements. Subsystem design. Mission analysis. Functions analysis. Functions allocation. Environmental and skills analysis. Manpower and training analysis. Systems design and implementation. Operational management. Implications of a human-factors design approach in the human-service industry.

PART II. Human Services and the Public Administrator

3. The Evaluation of Human Services and Their Interdependencies

   Human-service systems as they are today. Explicit developmental policies. Surface traffic control systems. Air traffic control systems. Fire protection control systems. Law enforcement and criminal justice systems. Sanitation service systems. Health-care service systems. Mental health service systems. Public welfare service systems. Human factors and the "Quick-Fix" Era.

functions. Systems evaluation: Cohort gathering, living skills indoctrination, vocational skills indoctrination, budgetary sustenance. A final system note.

List of Figures

11

List of Tables

List of Exhibits

Preface

After thirteen years in human-factors military hardware systems analysis (i.e., targeting missiles, Extra-Vehicular Activities space missions, Vertical Takeoff and Landing pilot missions and simulation, etc.), I was presented with a developmental opportunity to apply these same aerospace methods to domestic systems. The development concerned federally supported systems research in community mental health—design for the delivery of precare, residential, and aftercare services in the community to prevent hospitalization and reduce chronicity of mentally ill patients.

William Eicker, the Principal Investigator and a clinical psychologist, had worked on the California delinquency studies with Aerojet General and Serendipity Associates. In this first and only encounter with the human-factors type of expertise, he found major potential for analytic penetration into the maze of tradition-bound mental health operations. As we worked together more and more the multiplicity of hardcore problems of the mentally ill were seen to require a more total system of human services. This was the direction our analyses began to take, while designated service functions were progressively subjected to a study of operational alternatives for support of behavioral and distressed deviants in the community.

Eicker began his work at the Adolf Meyer Center, a regional facility of the Illinois Department of Mental Health in Decatur, Illinois, in 1966. He left in 1969, after three years of development work, to undertake a similar project at Brandeis University in Boston, while I remained to continue the systems work we had begun in the eighteen-county region of east-central Illinois. The human-factors analyses then began to consist of more detailed functions, situations, and task analyses in supportive health, welfare, criminal processing. etc., research for various classes of target populations. In fact, we then began to work more extensively with other agencies in the system complex, such as health-care councils, the Office of Economic Opportunity, Headstart programs and school districts, human relations councils, youth and delinquency offices, community development and housing, etc.

When working in these domestic fields, one must increasingly realize the system implications of any domestic problems—the hardship and burdens imposed when harassing poor relatives to pay for the treatment of their mentally ill next of kin, the disruptive influences of commitment practices on families, the perpetuation and reinforcement of poverty through unstudied welfare policies; in fact, an expedient, short-term nature of solutions is evident for any human service function that operates as an isolated, fragmented nonsystem component whether it be a mental hospital, a police precinct, a hook-and-ladder company, an isolated housing development, a new roadway, plastic garbage bags, or welfare case work. Such fragmented, unstudied nonsystem components can only repeat the same mistakes, create the same problems, and reinforce the same errors that have been operating for decades. The dysfunctional nature of a great many of our institutional and human-service functions must eventually be confronted if we squarely and honestly would seek to improve our life quality as a total society, or, indeed, even to sustain it.

Working in the human-service area of mental health provides abundant opportunities to observe the sociopolitical impediments that operate when attempting to develop or improve service systems; yet, at the same time, creative opportunities are evident, as well as challenges for those with the technical expertise to improve the human condition.

A human-service network might be developed under the cognizance or orientation of any central service function, as is frequently the case in mental health. Thus, human-service nomenclatures are found under the umbrellas of mental health, children-and-family services, juvenile delinquency, etc. Though my major human-service background in applying human-factors technology has been under the rubric of *mental health,* a human-service system, particularly for purposes of human-factors systems analysis, must assume more inclusive dimensions. This is the approach that has been taken in the present text.

The increasing scope of interest and development planning in the organized body of human-factors expertise lends major support to a broad definition of human services. In the past decade human-factors conference content has been increasingly expanded in the topical areas of paper sessions concerned with domestic problems. Since 1970, for example, topics of the Human Factors Society symposia and sessions have included commercial and municipal information systems, transportation and mass transit, environmental and architectural design, urban planning and social change, consumerism and product design, pollution control, design for disaster, health care and preventive medicine, etc. Indeed the overwhelming character of contemporary problems of pollution, traffic and urban congestion, undisposable solid waste, mass education, health care, crime control, ad infinitum, virtually demands the attention of the best scientific and analytic talents available. Many of the human-factors ilk are already becoming involved, and are increasingly to be found doing analyses of urban housing and development, city fire departments, health services, educational techniques in the schools, design for the developmentally disabled, urban transportation and criminal justice, to name but a few.

The aerospace technology is potentially available to advance almost any area of human services; though indeed to prove its worth, it must also become increasingly sensitive to the evaluation of human services, their effectiveness and operational costs. There is, in fact, an increasing urgency to develop more meaningful system measures to monitor our quality of life. Perhaps it is even necessary to reassess our total value system, if we mean to progress in conquering famine, mental illness and poverty, gross inefficiencies and inequities in our urban and business lives, and indeed to alleviate the stresses that lead to the early death of many of the most productive members of our society.

Development of the present text is directed to the reader whose basic interest and problems reside in public and social service—penologists, social workers, recreationists, police commissioners, fire commissioners, welfare commissioners, city managers, mayors, legislators, personnel managers, political scientists and educators in various of the fields which deal educationally with the professionalizing of these services. Substantively, major treatment is given housing, mass transit, crime and law enforcement, health and mental health, and public welfare, recreation, etc.; but in principle any similar public services may be comparatively addressed.

Urban crises in the recent past have revolved about racial tensions, transportation, police protection, etc.; and now funding of inner-city services has become a critical urban problem. These may often be directly attributed to deficiencies in the design, management, and administration of such services in reacting to crucial public issues in service priorities. It is the intent of this book to provide a description of the methods that have proven successful in the engineering of other such complex systems, and could, for the conscientious public official, contribute to the success of his efforts at systems construction and management.

J. H. Burgess
Monticello, Illinois

Acknowledgments

The author wishes to acknowledge, with gratitude, the inspiration and assistance provided by the countless professionals, as well as recipients of human services about which this book is concerned.

Dr. William Eicker provided many aggressive and inspiring insights in mental health, and an insistence on systems approaches new to the industry, as did Harold Halpert of the Systems Division in the National Institute of Mental Health.

Dr. Robert Wallhaus, Dr. Ronald Nelson, Kim Greider, and Dan Nellis also lent their expertise and encouragement in developing human service analytic principles. A multiplicity of nurses, social workers, community organizers, psychiatrists, educators, and human service agency personnel too numerous to mention augmented my human-factors background with much of the substantive content of the book.

Major credit is also due Eugene Ruskin, the Staff Librarian at the Adolf Meyer Center, who fed me innumerable reference documents in preparation of the manuscript.

21

System Design Approaches
to Public Services

Part I

A Systems Perspective in the Modern World

A systems outlook is often difficult to achieve, since it involves projecting outcomes and consequences over time, and seems to involve more than immediate problems and circumstances. However, without a great deal of imagination, one might realize that sending a convicted head of the household to prison will increase the welfare case load. Welfare systems predicated on the Elizabethan Poor Laws, or purely a dole principle, will reinforce and perpetuate poverty through the assumption of a "bootstrap" capability and initiatives simply not found among the poor. A broad systems view examines these inter-dependencies as they are found to occur in nature and in man-made environments. Their importance in public service design and management is highlighted to demonstrate what could be done if the vicious cycle of nonsystems practices is to be overcome—if vested interests, fixity of professional intent, and bureaucratic staticism are to be finally exposed as constituting a major block against the improvement of human services.

1

Introduction

The Importance of a Systems Approach

The classical human-factors system approach to design-criteria analysis is predicated upon the essential interrelationships that operate among men and physical components when performing extraordinarily complex missions. A total weapon system, for example, is built around a basic war policy with specified conditions of destruction or potential kill requirements. This must involve the weapon, the delivery system, the support equipment, personnel subsystems, training equipment, brick and mortar sites, logistics and maintenance equipment, etc., all to be designed to operate within required tolerances of time, reliability, and accuracy at theoretically minimal costs.

Yet, systems do not occur simply because they are fancied or viewed in such a fashion; it becomes, rather, a pragmatic fact that such complex interdependencies occur necessarily within nature; and to recognize this fact is to assume increasing control over conditions and operations for desired outcomes.

SYSTEMS MOST EASILY IDENTIFIED IN NATURE

Current crises in overpopulation, food and energy supplies, pollution and environment, etc., have served to heighten our awareness of ecology and the sensitive system balances of nature. In fact, upon reflection it can be seen that virtually nothing stands as an independent entity, but exists, operates, and manifests itself only in relation to other things. Several examples drawn from astronomy, meteorology, and biology may serve to illustrate these essential systems of interrelationships found in nature.

Astronomy

Someone may have first imagined that a few stars resembled a dipper in their arrangement. The image of a dipper served both to keep the arrangement in mind and to describe a grouping of stars. Since this group of stars maintained its arrangement from night to night, year after year, the grouping displayed a systematic simple spatial pattern in nature. The Big Dipper may be considered a system by virtue of the fact that it is not seen as seven separate stars but as a group of seven stars together. Similarly, the earth, moon, other planets, and the sun have been found to occur together, not only spatially, but as bodies functionally related and pulling on one another. By the same process of association, in the study of functional relations, systems may be designated as planetary-moon systems, mountain systems, river systems, etc. All are spatially if not functionally related.

The Big Dipper may be regarded perceptually as a system of stars, but the stars of the Dipper are functionally a part of a general host or cluster of stars, which are in turn part of a vast pattern of cosmic dust that make up a galaxy within a system of galaxies. Within the galactic system, motion components of star clusters whirl about a nucleus swinging in and out within a galactic arm. Thus, the solar system in motion may cut through galactic dust fields that move from the dense center of the galaxy to the attenuated dust fields of the outer arm. Dust fields of greater or lesser density trap energy within the solar system and its planets, literally moving mountains on earth within the system of influence.

Meteorology

Gases generated from early chemical decomposition in the formation of the earth created an atmosphere of predominately oxygen and hydrogen; this constitutes the basic energy system of the earth which is subject to systematic variations in ice formation, winds, rains, and storms. Systems to which the atmosphere is subject of course include the generation of further gases and contaminants from within the earth environment, the system of solar-galactic interactions, and variables within the atmospheric system itself.

Solar-Galactic System—as the solar system whirls inbound to the dense galactic nucleus of cosmic dust, heat becomes trapped within the atmosphere from the increasing density of the dust. Ice caps on earth melt, thus flooding coastal cities, while the earth's population of animals and men migrate to high land areas as has occurred in the eras of geological time.

An Atmospheric System—a diagram of a wind system at the seashore is illustrated below:

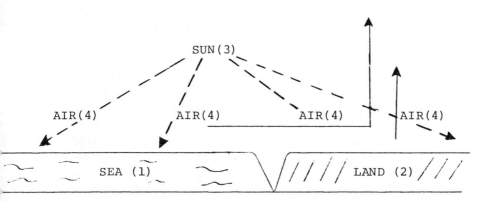

The system includes: (1) sea, (2) land, (3) sun, and (4) air. When the sun shines on both sea and land, the land warms up more than the water. As the land becomes warmer than the water, the air over the land becomes warmer than the air over the water and rises. The air over the water moves toward land to take the place of the air that is rising over the land. This system of air moving in from the sea is the sea breeze blowing on shore, and illustrates the complex interplay of atmospheric variables.

Biology

Systems in biology are perhaps the most dramatic in illustrating the interplay of interdependent components. Out of the chemistry of the earth, organic chemical matter evolved into those life forms which make up the systems and subsystems of living things. Regularities in the life cycle of a fly, for example, the egg, then larva, then pupa, then adult fly, is a system. A larva is not a separate animal, nor are the egg and the pupa unrelated objects. Rather, together they make up a system. While an ovule in a flower is developing into a seed, the ovary is developing into fruit. Such concomitances are systems. Biological systems, as components in any context, are artificial when separated out and treated as individual entities. To be convinced of this, consider the grouping of stomach, intestines, etc., into a digestive system, and then question if the heart is part of the digestive system. The textbook answer would likely be no. Yet, for how long would the stomach digest food if the heart did not supply it with blood? Therefore, with justification the heart might be included in the digestive system. This might also include the lungs, on the grounds that if the lungs did not get rid of the carbon dioxide produced by the stomach, intestines, and digestive glands, digestion of food would not take place.

Systems of organisms also include communities, societies of animals and plants, their growth, competition, and struggle for existence, both in the ecological and genetic sense. Human societies also occur in nature, with a system's interplay determining the growth of human populations, the emergence of international armament races and conflicts, the horrendous congestion in housing and transportation, etc., all of which may be described by equations similar to those used in ecology.

Sociology, with its allied fields, is essentially the study of natural human groups or systems ranging from small aggregates such as the family or work crew, through innumerable intermediate-size groups of informal and formal organizations, to the largest units that make up nations, power blocs, and international confederations.

MAN-MADE SYSTEMS

Man, in the ecological sense as a superinfluent, has evolved systems, often without design or intent, that run counter to functional design implications, often lacking in operational effectiveness presumed to meet implicit goals. Rather, those systems that purport to serve man and the human community may act adversely, inadvertently congesting, contaminating, destroying, distressing and undermining human societies; or those human-service systems that have, indeed, developed through civil, industrial, or social engineering efforts may be grossly deficient where more total systems implications have been neglected, vis-a-vis in failure to apply systems design discipline and human-factors considerations.

Man-made systems that have evolved in modern times, often without appreciable design or effective planning effort, may be seen as those ranging from the simple mundane problems of refuse disposal to the general health and welfare of the citizens of a community. Several systems implications of such man-made complexes may be illustrated as follows:

Garbage Disposal

Population growth and the manufacture and distribution of goods from which solid waste materials derive, such as old newspapers, packaging materials, inedibles and worn-out appliances, determine the current and projected rates or volumes of requir~d refuse disposal operations. The total system involves various alternative modes of handling, processing, disposal or recycling. Within the system must be considered noise, odor, general clutter, obstructions, degraded aesthetics, toxicity and health hazards, the lowly status and roles of garbage handlers and inappropriate human operator functions.

Police and Criminal Justice

Such a social-control system is extensively interactive and comprises multiple subsystems with myriad interactive elements. The criminal justice subsystem, for example, through arrests and convictions interacts with welfare, child care, and educational subsystems by creating new welfare cases in nonsupport, neglected children when the mother must work, high school dropouts when the adolescent loses self-esteem or must work to support the family, etc. The police require an extensive communication network and information base to determine the nature of misdemeanors, felonies, or ordinance violations, and identify habitual violators to be apprehended. Judicial functions inordinately delay adjudication and overload individual judges who may be inappropriately performing functions more optimally suited for machine processing. The correction system is inadvertently fostering and perpetuating criminal recidivism, while society at large continues to generate first offenders among the relatives and offspring of the recidivists, the impoverished population, and through the closed nature of its self-perpetuating criminal justice system.

Fire Protection

The rage of uncontrolled fire has plagued human and animal life since prehistoric times. Forest fires still leave vast gutted areas, killing all life in their consuming conflagration. The man-made structures and properties of civilized man are no less subject to the onslaught of a terrifying smoking and blazing holocaust. Man's modern systems of fire control must include prevention, early detection, and rapid efficient response. A fire-control complex, in all its system's ramifications, includes structures and products designed for fireproofing, the sniffing and finding of fires before they become uncontrollable, maintaining standby and ready status of the entire fire-fighting apparatus—vehicles, equipment, personnel subsystem, skills, etc., the prevention of boredom and loss of skills in the human components from prolonged idleness in waiting, rapid deployment of effective extinguishing devices with a research arm constantly improving these, survival and lifesaving equipment, safety and inhalation devices to minimize injury to victims and service personnel, while minimizing costs and damages.

Transportation and Vehicle Systems

The automobile, bus, truck, monorail, subway, train, or airplane is not bounded simply by parameters of man-vehicle machine interactive control. The complete system of interactive elements of a transportation system includes roadways, railways, airways, terminals, access, flow, safety, and every aspect in the total transport complex for materials and personnel. A new subway subsystem, for example, might lighten the flow in surface transportation but introduce points of congestion in pedestrian traffic requiring access control and regulation of flow. Gasoline station and parking-lot activities would fall off, etc.

Air transportation systems are also more than airplanes. They include baggage manufacture, merchandising, distribution, transportation to airports, passenger terminal traffic and baggage control, terminal communication, air-to-ground communication, automobile parking, ticketing control and reservations, food preparation and airborne serving, passenger comfort and safety, blind landing radar and air traffic control, pilot training and maintenance, stewardess schools and standards of practice, public relations and advertising, aircraft research and manufacture, etc.

Automobiles are more than internal combustion machines with transmission, wheels and styling. They include a system of roads, spare parts, gasoline and oil production and distribution, traffic signs and road paint, signal lights and casualty insurance, parking lots and driver licensing, traffic courts and driver training, factory workers and automated production equipment, economics of the profit motive, merchandising, pollution, etc.

Health-Care System

Health-care systems comprise a vast complex of primary, secondary, and tertiary prevention programs and individual service delivery components. Interactive elements of all programs relate to a multiplicity of other health and welfare components. Death-control health programs, for example, may create housing, feeding sanitation, and mental health problems through increased life expectancy and general population increase. Conversely, birth-control programs may interact on welfare, unemployment and poverty, and mental health through the alleviation of associated population pressures. Introduction of medical preventive programs, such as the Salk vaccine for polio, may have major impact on rates of disability, crippled children and adults, unemployment, welfare rates, etc.

General health hazards also have multiple ramifications. The widespread habit of cigarette smoking, for example, may be called a "cigarette-smoking system." This is thus not merely a tobacco-burning, smoke-spewing, coughing specimen of modern man. It is tobacco growing, labor problems, merchandising, television and radio messages (or their prohibition), printing and photography, cancer research and public health, advertising and public relations, packaging and marketing, economics and legislative lobbying, etc.

A hospital system as an individual service-delivery component is also more than beds, patients, nurses and charts. It is rather a large busy integrated complex of equipment, staff, and information. The hospital system includes business, accounting, logistics, clinics, medical data processing, training, treatment, patients, interns, doctors, nurses, orderlies, towels, medicine, x-rays, accreditation, radiation, etc. A systems approach to hospital operations may attempt to establish objectives for improving hospital functions and projected parameters— number of patients to be accommodated, kinds of illnesses, examinations, data processing, accounting, traffic control, visitor control, etc. Comparative cost and performance analyses may also be involved for possible alternative designs of future hospital systems.

Mental Health

Mental health is perhaps one of the most equivocal of man's contrived service systems; it bears marked kinship to the police and criminal justice systems in that deviant and disturbed behaviors are seen basically as problems of social control. The evolved system is one of vast complexes of mental hospitals, elaborate interactions of psychiatrists, psychologists, nurses, social workers, and various levels of subprofessional personnel. With current emphasis on community treatment (following the Community Mental Health Act of 1963), the system has come to include community mental health clinics, private physicians, general hospital psychiatric wards, community group care homes, sheltered workshops, etc. Within the complex is an assortment of subsystems, some of which are missions directed toward outcome goals, such as returning geriatric patients to the community, or developing socially self-supporting aggregates of chronic mental patients in the community. The total system implications of mental health services, however, are of necessity tied in with poverty, a general health and welfare concern for the human condition, and the myriad of self-fulfilling prophecies and abuses that men and their societies have unknowingly perpetrated on each other since the Middle Ages.

Welfare Services

Welfare service systems are notoriously fragmented, serving only bits and pieces of human needs. The public aid component evolved from the English poor laws calculated to maintain the impoverished at a subrebellion survival level. Out of deficiencies in the welfare service system, operating without design goals or stated service intentions, come child abuse, the disinclination to engage in self-supporting work, the perpetuation of subproductive work skills, criminal enterprises, mental illness, homicides, suicides, alcoholism, drug addiction, and other forms of uncontrolled self-indulgence.

SYSTEMS AND HUMAN-SERVICE ENGINEERING

The human-factors systems engineer and his multidisciplinary counterparts, in the several decades since World War II, have diligently developed analytical approaches that indeed lend themselves to the gamut of human-service design problems. Actual applications to non-military operations have only recently begun to evidence themselves in other than theoretical approaches. The actual, vivid and concrete input of human-factors engineering to human-service industries is now becoming a reality as part of the continuing systems and industrial revolution of our current era.

Systems engineering and its man-machine components now make available the nucleus of the new technology. Early man-made systems concerned simplex power engineering or the simple release of energy in steam or electric machines, from the simple light bulb or steam engine thence leading to today's complex computers and automated machinery. Self-controlling machines, with a complexity as simple as that of the thermostat or as complex as that of the fantastic self-steering missile, are among today's complex man-made engineering systems. Simple early designs for machine applications are now being expanded to include all elements of machine interaction. Design of a steam engine, an automobile, or a radio receiver was once within the competence of a single engineer trained in one specialty. Complex ballistic missiles or space vehicles, however, can only be designed, assembled, and integrated from components originating in such heterogeneous technologies as mechanical, electronic, and chemical engineering and the engineering of linkages between man and the machines he operates as part of the total systems complex; innumerable financial, economic, social and political problems are also necessarily involved in the design. Automobiles and their traffic are not just an aggregate of vehicles in operation, but require a vast and intricate control system involving myriad traffic situations.

When system objectives are specified, development of ways and means for implementation requires systems specialists and teams of engineers to consider alternative solutions, and to choose those most promising for efficiency and minimal cost in a tremendously complex network of interactions. Elaborate techniques and computers must often be used when solving the complex design problems that now far exceed the capacity of any individual designer. The hardware of computers, automation, and cybernation and the software of systems science make up this new technology which has been called the "Second Industrial Revolution."

ANALYTICAL METHODS APPLIED TO HUMAN-SERVICE SYSTEMS

The systems commitment to human services must issue from stated public policy and nonpartisan agreement within the political arena, or where otherwise vested interests operate. Given such a firm policy, without contamination by extraneous interests, goals may then be explicated and from these the *systems operational requirements* derived. In order to reduce or eradicate poverty within a specified time period, for example, it would first require that exhaustive operations research analyses be undertaken to relate the multiple interactive system factors of the poverty condition. Systems design would then develop from the multiple service requirements and *subsystem components* essential to accomplishing the total system goals, e.g., sustenance, family management, counseling and vocational training, job placement, and maintenance of work morale, etc. If these are considered the subsystems, then a subsequent *functions analysis* would establish the essential provisions or broadly conceived operations that must be performed within and among all of the subsystems for overall systems performance. Sustenance, for example, would require that food and living accommodations be provided, yet in keeping with total system goals sufficient to maintain good morale, while discouraging dependency, to culminate in self-sufficiency.

From such functions may be derived the means by which the desired outcome is to be accomplished. This process is classically referred to as *functions allocation,* wherein machines may be employed to distribute financial allotment, or a total sustentative situation may be provided without the exchange of money. Such functions are established on the basis of *trade-off* advantages in terms of costs or effectiveness in bringing about system goals.

When functions have been designated, a *task analysis* of the human operators or workers within the delivery system may then be determined following a *man-machine* functions allocation.

The human performance component of human-service systems often involves only man-to-man relations as direct service systems are currently designed. Machine or hardware components are perhaps more in evidence in such systems as sanitation and environmental protection, fire fighting, etc. However, a vast array of potentially new design features in other facets of service may also be potentially redesigned, as in computerizing the criminal justice system, or automating mental health therapy through the principle of the teaching machine. Such applications must await more precise trade-off analyses of advantages and effectiveness in accomplishing systems goals.

The task analysis involves direct interfaces of service operators, specifying the detailed interpersonal, or possibly man-machine, interactions necessary. The social worker, for example, may be required to relate to an impoverished individual in such a way as to encourage his full participation; or the task may involve interfacing with a computer to obtain information and service access to meet the individual's requirements. These will require definitive skills to be derived as specifications from the task analysis; from the task analysis, *qualitative and quantitative personnel requirements* may be established, and required *training functions* determined. The entire systems analytical process is a requirements-oriented one. Necessary functions are determined only as these are instrumentally related to desired outcomes. Throughout the entire process, alternative means of accomplishing these functions are under constant scrutiny, with modifications and redesign introduced when contributing to reduced costs and/or improved effectiveness.

DESIGN AND IMPLEMENTATION

The foregoing presents the classical analytical approach to a human service system development. The systems now operating in established entrenched human-service delivery modes, however, are not easily preempted, though costly and notoriously ineffective. Political interests, unyielding tradition, and vested credentialism of the controlling professional factions, of course, militate against sound rational system approaches to design, and perhaps present formidable barriers to implementation. Progress with the human-factors systems approach to human services may ultimately lie, as it once did with military operations, in the alarming realization that man cannot perform in a system where no design consideration has been given to his capabilities and limitations. Moreover, our cost-responsive society may become increasingly sensitized to the waste in dollars, manpower, and human dignity incurred when total system design parameters are neglected or blatantly obscured. Meanwhile, design fixes may be increasingly introduced in current system components,[1] while rare opportunities present themselves for complete subsystem design. The latter situation is illustrated in the nonresidential lodge of a mental health subsystem as described in chapter 12.

NOTE TO CHAPTER 1

1. Such developments are perhaps exemplified by the extensive driver-behavior studies found in the human-factors literature, which in turn provide the basis for design fixes in the automobile-road system.

2

Human-Service Systems
and Human-Factors Methods of Analysis

The basic description of a system under design must begin essentially with a statement of policy. Proceeding from the policy formulation, rational determinations are made to specify unequivocal objectives, termed "System Operational Requirements" (SOR). These become the determining vectors of design criteria; all subsequent activity is then directed toward developments designed to meet the SOR. From the SOR, design specifications and subsystem requirements are then deduced and translated into mission operations, i.e., the detailing of sequential phases of operation. These in turn are refined as general functions with further break-out specified on the basis of man-machine functions allocation, equipment, facilities, and specific skill requirements, their availability or the necessary training to be instituted to develop these. The entire procedure consists of macro- through micro-analysis, with derivations made at successive levels of increasing detail. Logically related solutions are arrived at for requirements at each level, with the ultimate system design thoroughly consistent with stated policy. A total developmental sequence of the systems design process is outlined in the block diagram of Figure 1.

A HUMAN-SERVICES POLICY FORMULATION

The assumption of a consistent working policy is generally predicated on a single-point authority over system operational areas, which may be a reasonably tenable assumption in the military and industrial sectors. Human-service system design implicitly makes the same assumption, and the single-point authority is further assumed to be governmental managerial control that matches the boundaries of the system to be designed. Government, however, is not a single-point

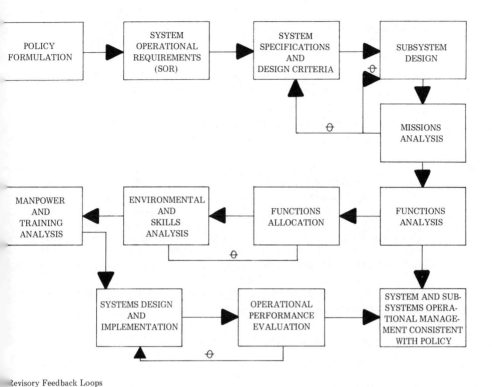

Revisory Feedback Loops

Fig. 1. Developmental Phases in a Human-Service System Design

authority, and in fact tends to be fragmented with multiple political levels and agencies each having discrete and separate jurisdiction over particular civil and social problems. Moreover, few government managers are conversant with, or cognizant of the need for, a systems approach. Most civil state and local governments make little allowance for long-range planning, and continuity is frequently lacking when offices and groups of elected officials, who are the decision makers and administrators, change radically during a systems development program. Incumbent bureaucratic inertia also presents an organizational and psychological climate better suited to resist the use of system design approaches than to adopt them.

Explicit policy statements on human services, interwoven as they often are with political and campaign commitments, are usually not available nor completely in evidence. In fact, public policy, as an issue, is of such equivocation that many contemporary scholars advocate a systematic and ongoing study of social needs and practices.[1]

However, while lack of explicit policy could block or impede the systems design process, an aggressive systems analyst may logically circumvent this. Implicit policy elements, though fragmented, may often be indicated from operational trends and evidential premises upon which major areas of political and public concern are based. Thus, in recent years repeated episodes of concern and alarm are in evidence over the expanding relief rolls and the skyrocketing public welfare expenditures.[2] The policy to be deduced from this evidential base, for example, may be phrased more or less explicitly as follows:

> Total reform of the three-century-old Elizabethan Poor Laws is needed. Programs must be tailored to the specific needs of the categorically impoverished. Quality of life for the impoverished must be measurably improved through specificity of support functions predicated on personal, social and occupational goals and a wide band of skills training and opportunity developments. Overall expenditures must be reduced through efficient use of resources and effective individual outcomes.

Such an explicated human-services policy then permits "System Operational Requirements" to be formulated.

SYSTEMS OPERATIONAL REQUIREMENTS (SOR)

The broad and inclusive statement of policy must be translated into fairly definitive operational implications for systems design. Such implications must relate to the improvement of the life style or individual life situations of the categorically impoverished—the aged, unwed mothers, unskilled fathers or heads of households, neglected and abused children, unemployable or troubled youth, the blind and disabled of all ages, as well as those in the often unclassifiable welter of distress directly attributable to meager resources.

In further establishing operational requirements information, operations research studies may provide essential data for the framework of design specifications. Major categories of poverty may be identified, and effective programming of resources determined.[3] Simply increasing welfare benefits, for example, has been shown to increase the expenditures for fatherless families, since fathers cannot compete with the level of support provided.[4] The poverty group most costly in crime has been operationally determined to be the young adult male. Operations research studies have also shown that greater costs are incurred in processing for eligibility determination than would likely be saved by simply serving those few who would otherwise be found ineligible.

Such ongoing operations research analytic and trade-off studies serve to determine, derive, and detail systems design criteria as illustrated in Figure 2.

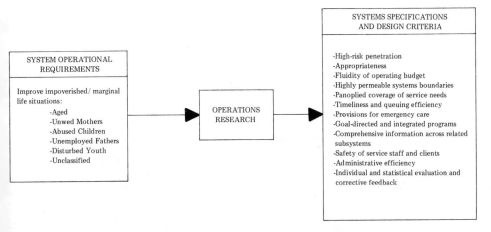

Fig. 2. Derivation of Systems Design Criteria

SUBSYSTEM DESIGN

Subsystem design may be developed as logical functionally grouped operations derived from systems specifications, and deduced from the nature of the "Input State" upon which the system is to act. This may be schematized as follows:

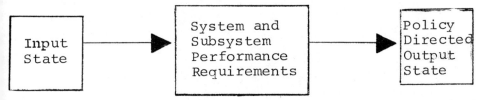

A human-services description of such an Input State, e.g., categorical and specific situations of impoverished individuals, is presented in Figure 3. A "template of trouble" is illustrated for which human services must be provided through a Screening-Dispatcher subsystem, to accomplish a desired "Output State." Figure 4 outlines a human-services network of subsystems based on a grouping of functional services designed for the template of troubles.

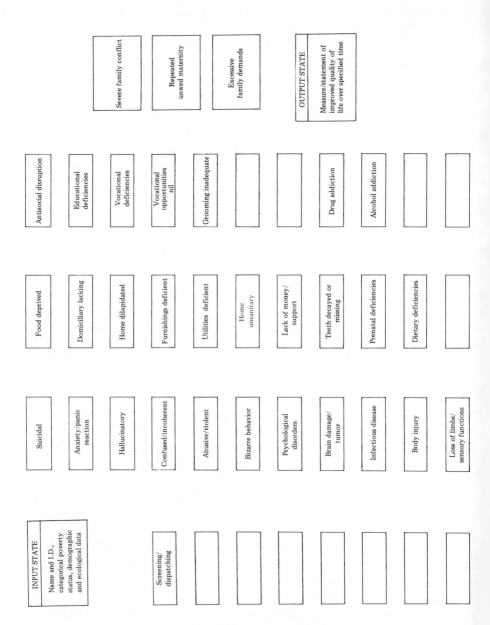

Fig. 3. Human-Services Screening-Dispatching Subsystem,
Template of Troubles

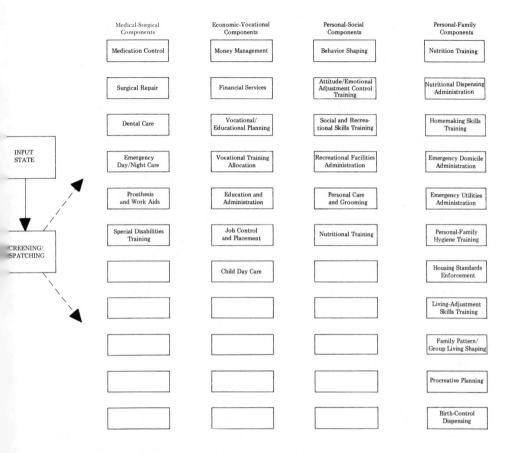

Fig. 4. Preliminary Human-Service Subsystem Design

An "Input State" of the total human-service system may be exemplified as follows:

White female under 25, abandoned by husband; two children, one *in utero*; depressed, one suicide attempt resulting in damaged trachea. Food and nutrition deficient; living in dilapidated housing with threat of eviction; no financial support; maternity diet deficient; children poorly fed and clothed. Education through grade school only; no occupational skills nor work experience. Teeth bad; physical appearance and grooming neglected. Socially and recreationally inactive and disinterested.

MISSION ANALYSIS

Employing the "Troubles Template" of Figure 3, specific characteristics of various Input States may be identified, i.e., isolated aged, disturbed youth, disabled, etc. The white female Input State, as described above, is plotted on the Trouble Template in Figure 5. A human-service mission may thus be analyzed by running such Input States through the system and its subsystems on a time line. Mary Jones, for example, requires Medical-Surgical treatment for her damaged trachea in the suicide attempt. She requires attitude-emotional adjustment control training. and social-recreational and grooming skills training in the *Personal-Social Adjustment Skills Subsystem*. She requires day care for her children, vocational planning and training in money management, and financial services in the *Economic-Voca-*

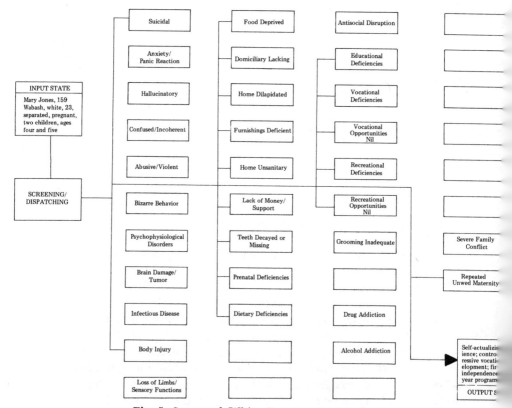

Fig. 5. Separated White Female, Trouble Profile

tional Adjustment Skills Subsystem. Emergency utility provisions, nutritional services, hygiene training and housing-standards enforcement are required in the *Personal-Family Maintenance Subsystem,* while family pattern shaping, procreative planning and birth-control services are indicated from the *Living Adjustment Skills Subsystem.*

Such a mission analysis of service programs must allow for proper phasing, and, within the time constraints established in the *Screening-Dispatching Subsystem,* must be designed for effective outcomes, while programming these to occur within specified time periods. As noted in Figure 1, during the course of missions analysis, feedback (Θ) may be provided for corrections of design specifications and subsystem design parameters as design deficiencies become apparent from operational evidence.

FUNCTIONS ANALYSIS

In the mission analytical process, specific functions required to accomplish outcomes may be identified. For example, the prescribing and dispensing of medication, or the cleaning of teeth and preparation of dentures may be seen as definitive functions in the *Medical-Surgical Correction Subsystem.* The training and care of children during the mother's absence, as well as vocational guidance and aptitude testing, may be seen as essential functions in the *Economic-Vocational Adjustment Skills Subsystem.* Functions analysis becomes a progressive detailing of required operations, regardless of how these are to be performed, whether to be automated or manually executed by highly trained professionals, subprofessionals, or specially trained technicians.

FUNCTIONS ALLOCATION

Traditional human-factors analyses require an assessment of functions to determine their suitability for machine processing. Thus, in engineering design, best estimates are made in allocating operations for machine design or for human operator performance through skills development. Agreement on what constitutes machine capability is of course predicated on current state-of-the-art, while machine capabilities are under continuing development. However, since basic human operator capabilities are fairly well known. in designing human-service functions for the human operator, several criteria may be employed. Inductive and management control functions are generally agreed to be best allocated to human operators, while routine, repetitive detailed clerical functions are already better performed by machines in the current state-of-the-art. Perceptual and pattern constancies remain a major machine deficiency, however, as do the machine's adaptability and its capability in filtering out noise, irrelevancies, and inaccuracies of the input. Table 1 presents several functional areas for which human operation in a human-service system may be poorly designed. In a *Screening-Dispatching Subsystem,* for example, initial personal human contact may be best indicated for continued recognition and rapport with familiar clients. However, specific processing functions may be best automated, such as input of demographic and ecological data for standarized service strategies. Disturbing personal and private affairs may indeed be best processed by a machine input function to avoid client embarrassment, hesitancy, or refusal to cooperate. Likewise for speed of processing, accuracy and assessment, aptitude and guidance testing, as well as training or teaching functions, may be best provided by machines. In any event, design and allocation of such human-service functions should be based on careful study of each operation and the required functional characteristics for man-machine trade-off in the design process.

TABLE 1

CHARACTERISTICS OF OPERATIONS POORLY DESIGNED FOR HUMAN OPERATION

Characteristic of Required Function	Human Operator Deficiency
Monitoring random events	Likely to miss details due to attention lapses and fatigue
Short-term memory	Recall unreliable due to distraction
Standardized coding and cataloging	Breakdown likely due to personal prejudices, vested interests, and individuated total client circumstances
Sustained high-level performance	Boredom and fatigue limits peak performance capability
Rapid data processing	Saturates quickly in number and duration of events-processing, resulting in delay of total process
Objective evaluation	Distortion likely due to individual expectancies and cognitive set
Immediate judgment and decision	Unreliable since delayed recall, review and deliberation are needed
Rapid data search	Unable to find coded detailed information quickly
Optimal standard strategy	Erratic and unpredictable

ENVIRONMENTAL AND SKILLS ANALYSIS

Following a functions-allocation analysis, the basic configuration of a human-service system, its subsystems and subsystem components begin to take shape. Here the design criteria of accessibility, timeliness, permeability of boundaries, etc., need be applied. Should such services be located ecologically within the high-risk areas, dispersed with satellites, or centrally aggregated? The vectors of systems design criteria, functions allocation, and feedback on effectiveness should provide necessary data to determine the required environmental settings of buildings, access ways, offices, computers, etc., and general layout of components, together with staffing and skill requirements.

MANPOWER AND TRAINING ANALYSIS

The personnel subsystem of a human-service system may be determined through continuing specification and refined descriptions of task details. Necessary skills may then be drawn from those of currently available professionals or subprofessionals to staff those positions compatible with traditional job specifications. Special training programs must otherwise be designed, with the necessary training functions determined, either for a totally new position or for task elements, depending upon the compatibility of available skills. As with the total system, training functions are subject to trade-off of costs with effectiveness, through on-the-job training or the use of ready-made courses, etc. Elaborate training program design may become excessively expensive, particularly when necessary skills may be derived through more standard training practices. Likewise, the use of highly paid professional personnel should be avoided, where personnel with less training could be employed who can be more specifically oriented to task requirements.

SYSTEMS DESIGN AND IMPLEMENTATION

Design and implementation of a human-service system seems completely amenable to standard methods of project management. The classical Program Planning and Budgeting (PPBS) approach in cost estimating and planning subsystem development target dates appears to be entirely feasible for use in the human-services implementation process. Documentation of design criteria, operations and human factors supportive research studies, of course, must be maintained throughout the design process. Other management methods in the systems development process are also equally applicable.[5]

The Performance Evaluation and Review Technique (PERT), familiar to every aerospace project engineer, may be applied in estimating time-to-completion of project elements. It may thus provide a means to focus project and resources on major problem areas and program lags, to manage and reallocate resources to adjust for proper phasing of subsystems in the total systems development process, and in controlling and integrating system components for proper phasing in arrival times.

OPERATIONAL PERFORMANCE EVALUATION

A process equivalent to the "Operational Suitability Testing" phase in weapon-systems development would also seem to be readily converted to the human-service systems context. This has traditionally been a kind of "shakedown" evaluation under operational conditions prior to making the system operational. In human-service operations, suitability testing might mean a pilot project designed for a small sector or community of population, to prove the value of a human-service system prior to extending its scope to the entire population in need. However, such pilot studies should already have served as documented supportive research for systems design. Rather, what appears needed in a human service system is ongoing evaluative feedback, flagging system deficiencies for improving outcomes in client-life situations.

OPERATIONAL MANAGEMENT

Management of an operational human-service system would likely constitute a style consistent with performing ongoing corrective action to improve results and reduce costs. A management procedure is illustrated in Figure 6. An information system or systems simulation model would provide baseline data for initial goal setting, e.g., "to reduce costs of the child day-care component by 10 percent in a fiscal period," or "increase productive opportunities for clients by 20 percent," etc. Systems models, such as that described by Burgess, et al.,[6] might be employed to determine deficiencies in the human-service system, while subsequent feedback in the baseline data would provide for signaling outcomes from program redesign.

The extensive use of computers and simulation models could also provide major benefit to operational systems managers. With valid and comprehensive systems models, executives and legislators may be

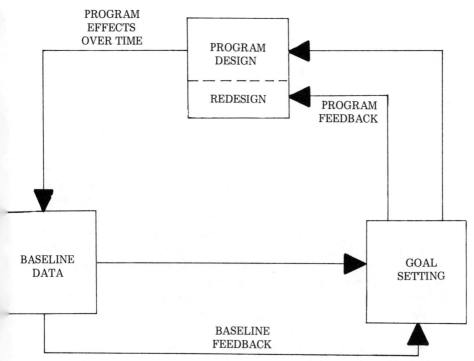

Fig. 6. Systems and Subsystems Management Paradigm. From John H. Burgess, "Mental Health Service Systems: Approaches to Evaluation," *American Journal of Community Psychology* 2, no. 1:88.

introduced to the wide-ranging implications of their legislation. Guaranteed annual income proposals, for example, may be tested in advance of actual measures through properly programmed models incorporating system correlates and algorithms. Building such models would necessitate incorporating multidimensions of the system, such as the effects of given amounts of guaranteed income on employment, health and nutrition, juvenile and adult crime, family unity, etc. Indeed, the best evidence available must be used to construct the models. If the death penalty were reinstated, for example, would it reduce the felonious crime rates? If parochial schools were funded by the government, would overall quality of education be improved? The operational management of a human-service system might draw heavily on such simulation models, maintaining an essential data bank that would readily lend. itself to model construction.

IMPLICATIONS OF A SYSTEMS ENGINEERING DESIGN APPROACH IN THE HUMAN-SERVICE INDUSTRY

The total human-factors systems design approach as described in the foregoing is of course presently only partially instrumentable. The current design situation for human-service systems is perhaps not unlike that of the aerospace industry during the 1950s. Only a fragmentary systems orientation was prevalent among aircraft designers. Then, complex mission requirements of high-velocity aerospace vehicles began increasingly to impose excessive workload and skill demands on human operators. In the resulting marginal performance and propensity for human error, project designers were required increasingly to turn to a human-factors systems orientation. This technology culminated in the sixties and early seventies in the completion of successful manned lunar missions.

A somewhat analogous situation appears to be developing on the domestic scene. The widely publicized crises in welfare, environmental pollution, energy resources, overpopulation, etc., increasingly give cause to legislators and social innovators to adopt the systems approach. The aerospace analogy, however, becomes somewhat tenuous in the face of the many bureaucratic and public resistance elements in public service systems. These, of course, were also present in the aerospace industry of the fifties, but of nowhere near the damning proportions that exist in human services.

The human-service industry today cuts across multijurisdictional boundaries which must first be integrated or consolidated if a systems orientation is to become reality. Extant traditional values and institutions must be readjusted if indeed a human-service system is to become possible. Such traditional institutions and practices as private investment, privacy and confidentiality, jurisdictional boundaries, free enterprise, etc., must be altered or grossly transformed in the course of a human-services systems development. Such alterations may include the following examples:

1. Extensive computer-teaching technology could result in a wholly centralized educational system, usurping traditional local school-district autonomy.
2. Social-systems data requirements for planning and operation of a human-service system will require release of such constraints as prohibiting invasion of privacy (a long-cherished value).
3. Long-range planning and social priorities may conflict with

laissez faire, and require reshifting of emphasis in the private-enterprise tradition.

4. Organizational power structures will be challenged and modified when information is to be freely exchanged in systems management. Long-standing power hierarchies, often maintained solely on the basis of the restricted and controlled flow of information, must yield to the demand in systems management technology for free and open exchange of information.

5. Vested professional, political, and academic interests must be subordinated to the accomplishment of system design objectives. Traditional service professions such as psychiatry, nursing, social work, occupational therapy, etc., must yield to the functional allocation of skills needed to accomplish subsystem and total system outcome goals.

The system analytical approach as described in the foregoing may be more briefly outlined as follows:

1. Assemble analytic and design group out of the governor's office, the office of the mayor, etc.
   a. Include system development and human-factors expertise in the design group.
   b. Obtain an assured funding source, under one jurisdictional head if possible.
2. Prepare or complete review of system requirements.
3. Develop feasibility studies or pilot projects.
   a. Evaluate alternatives.
   b. Prepare mission descriptions.
   c. Design and test feasibility service model.
4. Determine subgroups or subsystem elements.
   a. Complete functions analysis.
   b. Determine functions not feasibly automated.
5. Develop behavioral data on population served.
   a. Determine skills and aspirations.
   b. Determine opportunities and projected work availability.
   c. Determine resource requirements necessary to accomplish outcome desired.
6. Assemble or develop human-factors data for a serving subsystem.
   a. Determine environment and stress conditions, e.g., ghetto area climate, racial tensions, etc. Complete procedural requirements analysis, e.g., forms and applications to be completed, etc.
   b. Complete analysis of continuous control tasks, e.g., routine

work or training activities of client, and monitoring progress on status.

c. Complete information requirements analysis, i.e., concerning progress and problem areas.

d. Complete information flow and display formats required, e.g., monitor client on flow-through agency contacts, adverse service effects, etc.

e. Develop necessary job aids, e.g., routine computer printouts, terminals to mainline computer, etc.

f. Identify and design essential work stations, e.g., computer mainline centers, terminal stations at reception points, etc.

g. Determine maintainability and support requirements, behavioral and financial reinforcers, etc.

7. Complete time-line analysis, e.g., client agency services, durations and outcomes.

a. Identify workload requirements and capabilities, i.e., limitations of effective case-load demands, appropriate automated processes for client service elements, etc.

b. Determine tasks suitable for automation.

c. Determine coordination and team functions, e.g., group meetings to determine comprehensive client social, financial, etc., life-fulfilling requirements.

8. Determine operator skills and training requirements.

a. Develop task data.

b. Determine available skills to fulfill task requirements.

c. Establish areas of skill deficits and training or educative functions required.

d. Determine alternative approaches to education and training.

e. Develop and implement training and educational programs.

9. Evaluate and integrate system requirements.

a. Evaluate preliminary service program concepts and the theory and rationale.

b. Develop pilot projects in time-limited program.

c. Evaluate operational outcomes on a cost effective basis.

d. Determine long-term outcome and implications for operational evaluation.

Obtaining a single-point authority to whom to relate project goals and development is perhaps the chief difficulty in getting system programs underway. Failing this, the multijurisdictional heads of agencies or assemblages of agencies may become the separate bodies in management to whom a systems perspective may be imparted. The task of gathering such a body of separate decision makers in authority may fall primarily upon the systems specialists and community organizers, the behavioral scientists and community leaders, who might then bring all factions together including the voting public, to accept and undertake the systems project.

NOTES TO CHAPTER 2

1. I. Hoos, *Systems Analysis in Public Policy. A Critique* (Berkeley, Calif.: University of California Press, 1972) ; H. Lasswell, "Do We Need Social Observatories?", *Saturday Review*, August 5, 1967, pp. 49–52; L. Lessing, "Science Takes a Closer Look at Man," *Fortune*, January 1970, pp. 113ff.; J. Sawyer, "Does Military Psychology Promote Human Welfare?", *The Society for the Psychological Study of Social Issues Newsletter*, April 1970, pp. 3, 9.

2. R. Theobald, *The Guaranteed Income* (New York: Doubleday & Company, Inc., 1967) ; "Welfare Time for Reform," *Saturday Review*, May 23, 1970, p. 19; "The Welfare Mess Needs Reform," *Life*, July 1970, p. 28; "Runaway Rise in Welfare Spending," *U.S. News & World Report*, January 3, 1972, p. 43; B. Porter, "Welfare Won't Work, But What Will?", *Saturday Review*, June 3, 1972, pp. 48–52; R. Morris, "Welfare Reform 1973: The Social Services Dimension," *Science*, August 10, 1973, pp. 512–22.

3. "Class for Unemployed Stresses Little Things that Help Hold Job," *New York Times*, August 26, 1967; C. Wright et al., Prenatal-Postnatal Intervention: A Description and Discussion of Preliminary Findings of a Home Visit Program Supplying Cognitive, Nutritional and Health Information to Disadvantaged Homes" (Syracuse, N.Y.,: Syracuse University, paper presented at the annual meeting of the American Psychological Association, Miami, Florida, 1970) ; "Effects of the Earnings Exemption Provision on AFDC Recipients," *Welfare in Review*, January–February 1971, pp. 18–20; J. Rosenthal, "Female-Headed Families," *Chicago Tribune*, August 1, 1971; M. Rein and B. Wishnov, "Patterns of Work and Welfare in AFDC," *Welfare in Review*, November–December 1971, pp. 7–12.

4. F. Pivon and R. Cloward, *Regulating the Poor* (New York: Pantheon Books, 1971) , pp. 8–21.

5. H. Goodwin, "Improvements must be Managed," *Journal of Industrial Engineering*, November 1969, pp. 638–44; J. Hage and M. Aiker, "Routine Technology, Social Structure and Organizational Goals," *Administrative Science Quarterly* 14, no. 3 (1969) :366–76; R. Lindberg, "Handy Guide for Artful Planning," *Management Services*, January 1970, pp. 31–36; J. Walsh, "Technological Innovation. New Study Sponsored by NSF Takes Socioeconomic Management Factors into Account," *Science* 180 (1973) :846–47; M. Henry, "Research Utilization: A Problem in Goal Setting. What is the Question?", *American Journal of Public Health* 63, no. 5 (1973) :377–78.

6. J. Burgess, R. Nelson, and R. Wallhaus, "Network Analysis as Method for the Evaluation of Service Delivery Systems," *Community Mental Health Journal*, July 1974.

Part II

Human Services
and the Public Administrator

Human services have only infrequently been managed to produce a desired outcome. They have usually evolved helter-skelter—their organization following the technology, often lagging behind so badly that they never catch up with the world of technology; or to apply the technology with an intelligent foresight and planning skill, of which most public officials are completely capable. Public housing, roadways, surface mass transit, criminality and corrections, the poor laws and poverty, economics, education, mental illness are all capable of being hand-led with vastly greater effectiveness than is now being done. Without being overly optimistic, it indeed seems entirely feasible that through a grasp of the systems approach, and some measure of sophisticated systems management, those who now control human-service systems could themselves bring this about.

The present decision makers, as well as those who do the actual serving, might be made to see the effects of their actions. Random services evolving from ages past might be converted to a modern operation, one of efficiency, economy, and, above all, effectiveness for the hapless victims of poorly run and managed human-service processes.

3

The Evolution of Human Services
and Their Interdependencies

In community life it must be constantly reiterated that no solitary human-service component operates independently. Rather, these are subsystems that interact within the total community system; no one subsystem may be addressed without considering the relationships with all other components. Thus, housing developments relate to the location of industrial and commercial operating sites and job and employment opportunities in the community, as they do to surface traffic control, fire protection, law enforcement, sanitation and health services. Table 2 presents a conceptual exercise in relating such subsystem interdependencies. A rationale is implicitly indicated when specifying the nature of the relationships as "direct (D)," "indirect (I)," or "none (N)." For example, surface traffic control is directly related to domiciliary and industrial sites by virtue of flow density to and from these concentrated areas. Surface traffic control is also directly related to fire protection and law enforcement, since this is a primary component of these subsystems in their getting to the places of fire or criminal violation. Such a substantiating rationale may be developed to support each of the "D's," "I's," or "N's" in the cells of the matrix. The further clarification of such interrelationships among community subsystems may be developed through these logical type exercises, and more precisely through formal operations-research analysis[1].

The natural evolution of human services may be seen to develop as the needs arise in a community of people living together. Throwing water on a fire or removing its oxygen was found to control or extinguish it as part of the necessary experience of the community. The next logical step was to stand by with containers of water, etc., in anticipating fires. Other human services might be seen to have similarly evolved, each being directly dependent upon the extant technology and the level of scientific sophistication. For example, the health

TABLE 2
COMMUNITY HUMAN-SERVICE SUBSYSTEM INTERDEPENDENCIES

	Domiciliary	Industry & Commerce	Surface Traffic	Air Traffic	Fire Protection	Law Enforcement & Crim. Justice	Sanitation	Health	Mental Health	Public Welfare
Domiciliary		D	D	I	D	D	D	D	I	I
Industry & Commerce	D		D	I	I	I	I	I	I	I
Surface Traffic	D	D		D	D	D	I	I	I	I
Air Traffic	N	I	D			N	D	D	N	N
Fire Protection	D	D	D	N		D	I	I	I	I
Law Enforcement & Criminal Justice	D	D	D	N	D		I	I	D	I
Sanitation	D	I	D	I	I	I		D	I	I
Health	D	I	I	N	N	I	D		I	I
Mental Health	I	I	I	N	I	D	I	D		D
Public Welfare	D	I	I	N	I	I	I	D	D	

D – Directly and Functionally Interdependent
I – Indirectly and Causatively Interdependent
N – No Ostensible Interrelationship

profession for many decades at the turn of the nineteenth century practiced bleeding to rid the body of the hypothetical poisons causing the illness. On several occasions President George Washington had been bled up to a half pint of blood by his doctors in diagnosing and treating a variety of illnesses.[2] Two of three doctors called to his bedside for what today might have been diagnosed as a minor streptococcus infection elected to bleed him of a quart of blood; this, as many historians see it, was the direct cause of his death. Had today's most fundamental principles of experimental design been available in the technology, deliberate bleeding as a cure would never have been so widely practiced as to result in the death of our first president.

Today the technology with respect to public welfare services has likewise barely evolved from the level of puritanical culture at the time of the early colonists under Governor John Winthrop in Massachusetts. Wealth is still seen as a blessing bestowed by God for hard work, while poverty is still implicitly regarded as God's punishment. Advances in public welfare services are thus today still hampered by these widespread notions and implicit traditions of God's intent.

In the years since the Second World War, the technology in such areas as electronics and cybernetics has indeed advanced so rapidly as to permit human-service systems, such as law enforcement or surface traffic control, to operate with increasing sophistication. On California's freeways, for example, traffic sensors relay alarm signals to a control station when traffic flow is retarded below a nominal rate. Helicopters are then dispatched to the locus of trouble, the traffic snarl is corrected, with normal traffic flow resumed within a matter of minutes. These selective advances in many areas of technology have thus greatly expedited or accelerated the evolutionary processes in several of the human-service systems or subsystems.

HUMAN-SERVICE SYSTEMS AS THEY ARE TODAY

While many human-service systems today have evolved with high levels of sophistication, many are often only randomly effective; none have been systematically studied for interrelated factors bearing on total community systems. For the most part, developmental policies tend to be neither well defined nor systems-oriented, with functional characteristics reflecting, in perspective, the evolutionary style of development. For example, the standard automobile steering wheel is vestigial from a machine design that required two-handed force application and high leverage ratios (cf. the aircraft joystick). The

TABLE 3
Human-Service Systems, Functional Characteristics, and Coupling Modes

Human-Service System and Component Description	Explicit Development Policies	Current Functional Characteristics	Coupling Modes
Surface Traffic Control—predominantly wheeled, motorized land vehicles on open-road network	None. The Transporation Dept. opened in 1967. Johnson administration advocated the furtherance of systems analysis. No development objectives specified in ground traffic control. Traffic congestion, accident rates and air pollution have become of public concern, with policies emerging. Energy crisis currently predominant.	Manual and automated starting-stopping, steering, velocity and environmental control of passenger stations. Signs, codes, and symbols to be interpreted for time-limited action.	Visual contact surface; road signs; instrument panels; manual. tracking and procedural controls. Enforced rules and regulations.
Air Traffic Control—winged air-burning and rocket-powered air vehicles on chartered imaginary airway networks	None. (See above) Workload and tasks of air traffic controllers have become of concern, as well as increasing terminal and airways congestion, accident rates, and air pollution. Major policy is influenced from energy crisis.	Manual and automated initiation of takeoff and landing. Rules, regulations, and codes governing cruise flight and expected time of arrival.	Visual, radar, radio voice and visual display signals. Maps, charts, manuals, and enforced rules and regulations. Automatic landing equipment requires pursuit-compensatory tracking or monitoring task.

TABLE 3

HUMAN-SERVICE SYSTEMS, FUNCTIONAL CHARACTERISTICS, AND COUPLING MODES (CONT'D.)

Human-Service System and Component Description	Explicit Development Policies	Current Functional Characteristics	Coupling Modes
Fire Protection Rapid response wheeled ground vehicles and rotary wing air vehicles, various fire-fighting tools, water, chemicals and resuscitating equipment	None. Concern is evident for high rates of uncontrolled fire damage. Emphasis frequently placed on prevention.	Manual and automatic alerting, with service personnel on standby status. Quick reaction time operationally deploying various fire-fighting tools.	Auditory alerting signals from visual or automatic sensing devices. Manual control of vehicles and fire targeting for water and chemicals extinguishing Automatic water release, etc.
Law Enforcement and Criminal Justice—detection, apprehension, and correction of malicious, antisocial acts of juveniles and adults.	None. Reducing "crime in streets" has been a major political appeal. Need for prison reform is frequently publicized through prison riots, etc. Delays in adjudication are also evidenced as major concern.	Manual and automatic detection, with service personnel on patrol and standby. Manual apprehension of violator. Judgments are made on nature and magnitude of violations. Guilt or blame is established manually, and correction is primarily custodial or incarcerative.	Documented rules and regulations. Computer data base maintained on offender's identity. Architectural design of corrections facilities for retention and control. Historical adjudications maintained as precedence for current fixing of blame.
Sanitation—collection, routing, and disposal of solid, gaseous, and liquid waste products.	None. Heightened concern is evident about pollution and life-endangering elements extant from random	Public or private collection of solid individual household and industrial waste. Fluid and gaseous waste	Manual lifting of individual containers into semi-automated truck handling equipment. Earth-moving

TABLE 3

HUMAN-SERVICE SYSTEMS, FUNCTIONAL CHARACTERISTICS, AND COUPLING MODES (CONT'D.)

Human-Service System and Component Description	Explicit Development Policies	Current Functional Characteristics	Coupling Modes
		ducted to air or pipelines, thence to processing plants, or liquids dumped raw into natural waters, cesspools, etc.	waste accumulations in careless disposal. Reclamation and recycling interest is also evident.	equipment, chemical test equipment, incinerator installations, etc.
Health Services—prevention of epidemic illness; medical and surgical repair of diseased or injured individuals.	None. Several bills and practices have been enacted including Medicare and Medicaid. The Public Health Service Act with amendments sets forth guidelines. The National Health Insurance and Health Maintenance Organization movements are in issue. Public concern is evident about increasing costs and dearth of health facilities and trained personnel.	Primarily private entrepreneurs provide and control services on a dyadic basis in office or resident facility. Pharmaceutical elements are prescribed by diagnosis, or surgical tools used to remove or repair organs and tissues.	Offices and hospitals employed as physician laboratory to treat manifestations of illness. Visual displays of x-rays, graphic and oscilloscopic displays; cutting and handling tools. Documented principles, practices and ethical rules. Manual and computerized administrative data for retrieval and processing.

TABLE 3

HUMAN-SERVICE SYSTEMS, FUNCTIONAL CHARACTERISTICS, AND COUPLING MODES (CONT'D.)

Human-Service System and Component Description	Explicit Development Policies	Current Functional Characteristics	Coupling Modes
Mental Health—removal, habilitation, and rehabilitation of individuals with adjustment and coping problems, and behavioral aberrations.	None. The Community Mental Health Act of 1963 provides guidelines and needed funding support for community services development. Alternatives are being considered to the use of state or county mental hospitals, but the medical model and public intolerance for nuisance cases tend to dominate.	Architecturally designed facilities as self-contained communities. Primarily custodial functions with minimal system regard of individual's life situation and developmental goals. Unwieldy self-perpetuating bureaucratic and entrepreneural administrative situation.	Primarily manual persuasive and physical control of individuals by high staff ratios. Documented definitions, rules, codes and practices for individual and group management, through group talk, drugs, restraints and isolation. Computer employed administrative process and accounting procedures.
Public Welfare—financial assistance to public cases who cannot provide for themselves.	None. Major concern about excessive expenditure is in evidence by politicians and middle income citizen groups. The Allied Services Act proposes more "umbrella" coverage of service needs to alleviate individuals' problems and life situations.	Manual and automated procedures to determine eligibility and administer financial disbursement. Monitoring due entitlement, and prosecuting cases receiving illegal allotments.	Primarily manual interpersonal contact of staff with clients. Rules, regulations and legal documents govern interfacing; computer processing and retrieval of data is also under increasing development.

availability of powered servomechanisms would, of course, **obviate** this design today.

Modes of coupling the system components for human operation are also often random and error prone. Table 3 outlines contemporary human-service systems, current development, functional characteristics, and coupling modes for human-factors consideration. Several of such systems may generally be seen readily to facilitate employment of system approaches. Such systems might include traffic, law enforcement, and fire-control systems. Other of the systems, such as criminal justice and corrections, mental health and public welfare, seem more by nature to be fragmented, perhaps for reasons of social tradition and the force of conventional treatment processes. Human-factors coupling modes in these latter systems of course have largely never been considered, or at best are studied only within the confines of a closed-systems approach dealing with internal interdependencies.

Explicit Developmental Policies

Virtually throughout the entire human-service system network there appear to be no explicit developmental policies operating either through legislation, referenda, or effective executive leadership. There are indeed multiple cries of alarm; extensive editorializing by the news media, and research studies in the scientific community also point up the growing needs and crises in various human service areas. It is from such fragmented sources that substantive material must ultimately be derived for policy formulation, which must of necessity, from a systems standpoint, be made the cornerstone of any human-service system development program.

In a democratic society it may also be fallacious intent to rely excessively on presumed wisdom of the citizenry.[3] Consumer participation in program or systems development may best be limited to policy input—and then only by an educated and informed consumer. Indeed, the consumer may emphatically realize that health and dental care are inadequate, or that traffic is horrendously congested and housing is dilapidated and isolated from other services. He would, however, not presume to tell the dentist how to fill teeth, nor a builder how and where to construct houses. Consumer involvement may therefore be best coordinated in consensus responses as a needs-input to the policy formulation. It then becomes the task of human-service system designers and human-factors engineers to implement the policy and optimize the human-service delivery systems.

Surface Traffic Control Systems

In recent years growing interest and concern have developed about the needs for improved surface travel, though definitive development policies are yet to emerge. In 1967, the federal government formed a comprehensive transportation department, and in 1973 a mass transit bill was signed into law.[4] Increasingly the public has been harassed by traffic snarls, delays, prohibitive parking expenses, and general inaccessibility to the inner city with the decline and increasing inadequacy of mass transit facilities.[5] This has also given cause for increasing involvement of the human factors profession.[6]

While extensive human-factors studies are indicated in this human-service area, the need for continuing system problem definitions is also becoming more apparent. Human factors efforts may be directed most resourcefully to those areas in the present surface traffic-control system most crucially at risk. Operations research studies, for example, have shown that high accident rates occur at intersections, and in dynamic road situations where perceptual judgments and decisions are called for on a critical time base. Human-factors studies may then relate to enactment and enforcement of critical traffic laws.[7] High arrest rates, for example, are currently indicated for speeding violations. Operational research studies indicate that greater potential pay-off in accident reduction would accrue in deploying police personnel more at intersections, lane-changing situations, etc.

The dynamics of driver behaviors have been given increasing attention by human-factors engineers in recent years in developing criteria for improving vehicle and road system design.[8]

However, more central human-engineering contributions for improving this and other human-service systems may be best oriented in conjunction with operations research studies in traffic control. These could point to the critical situations in operations where accidents, congestion, high costs, etc., occur. Human-factors studies focused in such a way on the currently operating system could thus bear more fruitfully on current system needs.

Air Traffic Control Systems

For the past several decades, the major effort of human-factors engineers indeed has been in the aviation industry, for Air Force or NASA applications. Unfortunately, civilian or human-service air traffic control systems have been given only token attention. In this area as well, a developmental policy is needed, anticipating critical airways control problems. This human-service area, perhaps more than others that are poorly defined, should be amenable to a thoroughgoing systems- and human-factors approach.

Fire Protection Control System

Though completely amenable to systems- and human-factors approaches, with easily definable design parameters, this service system area has also been sorely neglected in the human-factors profession. Again, the lack of a definitive developmental policy, and the fact that substantive operations research analyses in fire protection have been minimal, may be largely responsible for the failure in effective employment of human-factors systems specialists.

Law Enforcement and Criminal Justice System

The component of this system, of course receiving the most attention (and most easily amenable to the application of systems technology), is law enforcement. This also receives periodically renewed political attention, as well as aroused public concern and intimidation. The current law enforcement system lends itself splendidly to the use of sophisticated technology in communication, time-line analyses for quick reaction time, the rapid and accurate identification of offenders, criminal laboratory criteria development in establishing guilt, etc.[9] However, again human factors systems input to this human service system, even in a fragmented form, appears to be virtually nil.

The more major contemporary system issues, however, seem now to revolve more about the total criminal justice system[10] (Newman, 1967; Miller, 1970). Operational studies indicate that "trust" and "integrity" in law enforcement officials is correlated with low crime rates, as is speedy adjudication of criminal acts. Public opinion surveys and research studies indicate that correctional programs are grossly ineffective. Nearly everyone is personally apprehensive about the growth of crime and the inadequacy of the total criminal justice system; but the apprehension is vague, with only tentative solutions in perspective.[11] Criticism is repeatedly leveled at the corrections system, and indeed many specific criteria of design for rehabilitation are available.[12] Policy formulation is bogged down in the morass of tradition and confused standards, however. Even so, given only fragmented definitions of reform, the human factors and systems specialists could conceivably contribute most significantly to development programs.

Sanitation Service Systems

Except for theoretical formulations and rare opportunities for design analysis, the human-factors engineer, and even systems engineers, have had but minimal input to this area of human service. Here, too, while this human-service technology has advanced very slowly in recent decades,[13] by virtue of the clearly definable physical processes and sequences, it is admirably suited to a human-factors systems design approach. Moreover, with increasing concern about pollution problems, workable developmental policies are quite likely to emerge, for which the implementation by human-factors systems design skills may be most appropriate.

Health Service Systems

Awakening human-factors interest is evident in this human-service design area, with a special health-interest group operating in the Human Factors Society.[14] Though widespread systems and operations research interest is evident in this human-service area, developmental policy formulation is still severely problematic. For the foreseeable future, human-factors contributions would seem possible only in fragments or segments of the health service system.

Mental Health Service System

Though this service area would be most functionally subsumed under a more aggregated human-service network, its current operation is heavily coded with medical model precedence. As in the general health area, human factors and systems contributions may be limited to isolated components or circumscribed operations.

Public Welfare Service System

A hypothetical human-factors systems development of this system was dealt with in considerable detail in chapter 2. Here, too, until more definitive developmental policies emerge, the human-factors contribution may be limited only to improving human factors within processing tasks of the current system.

HUMAN FACTORS AND THE "QUICK-FIX ERA"

Human-factors participation in a total systems design effort is perhaps now feasible only in definitively designed and managed military and NASA aerospace system developments. In contemporary human-service system developments, complete system cognizance by human-factors engineers has been realized only in isolated instances. Rather, the design problems most likely to be presented to the human-factors engineer are those of human error, excessive fatigue or boredom, alleviating undue stress, etc. A human-factors engineer, for example, may be called in to determine why an oscilloscopic display of cardiovascular action, or x-ray images, etc., fail to provide sufficient visual cues of pathology; or he may be asked to lay out an office area or the arrangement of computer console components, etc. He may thus address only single components of what may be a more complete system, and only in the case of troubleshooting a critical human error may have occasion to run a systems study—then with only minimal design changes possible. Indeed, this was the way the human-factors engineer got his foot in the door in the early days of aerospace design. When a system failed to work, a "quick fix" was called for, and the engineering psychologist was frequently the one to troubleshoot it. As human-factors capabilities become more commonly known in human-service industries, here, too, perhaps will the human-factors engineer be called in, and the "quick-fix era" will again be upon us.

NOTES TO CHAPTER 3

1. W. Wheaton, "Operations Research for Metropolitan Planning," *AIP Journal*, November 1965, pp. 250–59; "In Our Opinion: The Trouble with Urban Technical Planning Is That We Can Hardly Identify the Key Problems. . . ," *International Science and Technology*, July 1965, p. 17; R. Lindberg, "Handy Guide for Artful Planners," *Management Services*, January 1970, pp. 31–36; K. Watt, "Will the Future be Shaped by Rational Policies?", *Saturday Review*, October 28, 1972, p. 76.

2. J. Tebbel, *George Washington's America* (New York: E. P. Dutton and Company, 1954).

3. I. Hoos, "Systems Techniques for Managing Society: A Critique," *Public Administrative Review* 33, no. 2 (March 1973):157–64.

4. "New Department of Transportation Opens Officially," *Human Factors Bulletin*, April 1967, pp. 1–2; "Nixon Signs Highway Bill; Mass Transit Aid Allowed," Associated Press, August 13, 1973.

5. W. Owen, *The Metropolitan Transportation Problem* (Washington, D.C.: The Brookings Institution, 1956); C. Sundberg and M. Ferar, "Design of Rapid Transit Equipment for the San Francisco Bay Area Rapid Transit System," *Human Factors Journal* 8, no. 4 (August 1966):339–46; W. Hamilton and D. Nance, "Systems Analysis of Urban Transportation," *Scientific American* 221, no. 1 (July 1969):19–27; J. Rae, *The Road and the Car in American Life* (Cambridge, Mass.: MIT Press, 1971); W. Owens, *The Accessible City* (Washington, D.C.: The Brookings Institution, 1971); R. Buel, *Dead End: The Automobile in Mass Transportation* (Englewood Cliffs, N.J.: Prentice Hall, Inc., 1972); "Transportation and

People," *Society* 10, no. 5 (July 1973) :14–43; P. Barnes, "Is BART Any Way to Run a Railroad? So-So Rapid Transit," *The New Republic*, September 1, 1973, pp. 18–20; "Transportation: Chicago, Dallas, Toledo," *Science News* 104 (1973) :170.

6. H. Parsons, "Human Factors Workshop in Highway Transportation," *Human Factors Bulletin*, March 1972, pp. 4f.; R. Mortimer, "Human Factors at the Highway Safety Research Institute," *Human Factors Bulletin* August 1967, p. 4; A. Lauer, *The Psychology of Driving: Factors of Traffic Enforcement* (Springfield, Ill.: Charles C. Thomas, 1972).

7. P. Fergenson, "The Relationship Between Information Processing and Driving Accident and Violation Records," *Human Factors Journal* 13, no. 2 (April 1971) : 173–76; R. Miller, "The Needs and Potential for Cooperation Between Human Factors Specialists and Lawyers in Research and Development of Automobile Accident Law," *Human Factors Journal* 14, no. 1 (February 1972) :25–34.

8. A. Torf and L. Duckstein, "A Methodology for the Determination of Driver Perceptual Latency in Car Following," *Human Factors Journal* 8, no. 5 (October 1966) :441–48; F. Matanzo and T. Rockwell, "Driving Performance Under Nighttime Conditions of Visual Degradation," *Human Factors Journal* 9, no. 5 (October 1967) :427–32; G. Barrett et al., "Feasibility of Studying Driver Reaction to Sudden Pedestrian Emergencies in an Automobile Simulator," *Human Factors Journal* 10, no. 1 (February 1968); B. Cole and B. Brown, "Specification of Road Traffic Signal Light Intensity," *Human Factors Journal* 10, no. 3 (June 1968) :245–54; "Subcommittee Hears of Baffling Signs," *Human Factors Bulletin*, May 1969, p. 3; B. Davies and J. Watts, "Preliminary Investigation of Movement Time Between Brake and Accelerator Pedals in Automobiles," *Human Factors Journal* 11, no. 4 (August 1969) :407–09; P. Olson and R. Thompson, "The Effect of Variable-Ratio Steering Gears on Driver Preference and Performance," *Human Factors Journal* 12, no. 6 (December 1970) :553–58; D. Gordon and T. Mast, "Driver Judgments in Overtaking and Passing," *Human Factors Journal* 12, no. 3 (June 1970) :341–46; G. Johansson and K. Rumar, "Drivers' Brake Reaction Times," *Human Factors Journal* 13, no. 1 (February 1971) :23–27; G. Robinson et al., "Visual Search by Automobile Drivers," *Human Factors Journal* 14, no. 5 (August 1972) :315–24; M. Ritchie, "Choice of Speed in Driving Through Curves as a Function of Advisory Speed and Curve Signs," *Human Factors Journal* 14, no. 6 (December 1972) :533–38; C. Ladan and T. Nelson, "Effects of Marker Type, Viewing Angle, and Vehicle Velocity on Perception of Traffic Markers in a Dynamic Viewing Situation," *Human Factors Journal* 15, no. 1 (February 1973) :9–16; R. Lucas et al., "Part-task Simulation Training of Driver's Passing Judgments," *Human Factors Journal* 15, no. 3 (June 1973) :269–74.

9. J. Kennedy, *Police Administration* (Springfield, Ill.: Charles C. Thomas, 1972).

10. J. Newman, "Cops, Courts, and Congress," *The New Republic*, March 18, 1967, pp. 16–20; A. Miller, "Science Challenges Law," *Behavioral Scientist* 113, no. 4 (1970) :585–93.

11. R. Martinson, "The Paradox of Prison Reform—I. The Dangerous Myth," *The New Republic*, April 1, 1972, pp. 23–25; R. Martinson, "The Paradox of Prison Reform—II. Can Corrections Correct?", *The New Republic*, April 8, 1972, pp. 13–15.

12. J. Mitford, *Kind and Usual Punishment* (New York: Alfred A. Knopf, Inc., 1973); K. Burkhart, *Women in Prison,* (New York: Doubleday and Co., Inc., 1973).

13. E. Steel, *Water Supply and Sewerage* (New York: McGraw-Hill Book Company, 1960); V. Ehlers, *Municipal and Rural Sanitation* (New York: McGraw-Hill

Book Company, 1965); O. Fanning, *Opportunities in Environmental Careers* (New York: Universal Publishing and Distributing Co., 1971); "Report on the Bottle Battle," *World Environment Newsletter*, March 13, 1973, p. 42.

14. H. Bender and G. Rowland, "Technical Group in Environmental Design," *Human Factors Bulletin*, January 1972, p. 3; *Human Factors Bulletin*, "NATO Symposium on Human Factors in Health Care and Health Technology," September 1973, p. 8; W. Richardson and D. Neuhauser, "First Question in Health Planning: Does the Public Know What It Wants, or Not?", *Modern Hospital*, May 1968, pp. 115 ff.; S. Garfield, "The Delivery of Medical Care," *Scientific American* 222, no. 4 (April 1970) :15–23.

4

The Human-Service Mission

Human-service missions in principle are not wholly unlike those of any man-machine system operation. Multiple interdependencies operate among system components, any one of which may result in a breakdown of the system, particularly where human operators are involved.

In the early 1950s, when human-engineering technology was just beginning to emerge from psychological testing and experimental psychology, many of the sophisticated and complex aerospace systems turned out to be decision-limited with only marginal operational capabilities. The Aeromedical Laboratories in Dayton, for example, in contract monitoring of the RASCAL, an air-to-surface missile, questioned the human operator's capability to control the launch sequences. Launch tasks had been assigned to the air observer bombardier on the B-47, whom they regarded to be the most heavily burdened worker on the plane. A human-factors input thus became a mandatory contractual item, not in forethought, but due to the evolved marginally prohibitive workload that left the human operator error prone, and the entire mission in jeopardy.

Unfortunately, many such human-factors inputs have occurred as afterthoughts in design for mission operations, as in the case of a long-legged air observer bombardier who resigned from the B-47 crew when he observed that he would lose his knees if required to eject through the restricted opening of the downward ejection hatch.

When human-factors engineers have been called in to run the necessary analyses in uncovering such marginal situations, typically systematic and comprehensive systems analysis methods have been applied. In the present chapter, an example of such methodology is described as applied to a human-service mission; in this case, the context was placement of geriatric mental hospital patients into community resident facilities in a situation where public management was hampered by multiple factions, each of whose jurisdictional responsibilities militated against accomplishing the mission.

71

HUMAN FACTORS ANALYTICAL METHODOLOGIES

When the context for analysis concerns ongoing operations, a human-factors systems analytical methodology that has become a more or less standard practice may be outlined as follows:

1. Defining the systems problem
2. Setting the system boundaries
3. Specifying mission outcome goals
4. Adapting and/or devising analytical methods
5. Identifying and documenting critical functional areas contributing to the system operational deficiencies
6. Developing and documenting quick-fix recommendations
7. Coordinating implementation
8. Completing follow-up evaluation

Defining the Systems Problem

During 1969, legislation was enacted in Illinois providing for the transfer of "nonpsychotic" elderly patients from state mental hospitals to appropriate resident situations in the community. This included the finding of such alternatives for elderly patients whose mental processes were impaired as a concomitant of advanced years and who might otherwise be committed to such mental institutions.

The total problem was thus to establish the machinery for examining and placing all such elderly or geriatric patients into appropriate group-care living facilities in the community. In the face of the well-worn traditional pathways to state mental hospitals, and the penchant of traditional state hospitals for retaining their patients, the machinery of transfer and developing community alternatives for the elderly was markedly impeded. Late in 1970, the legislation simply was not being implemented with dispatch or efficiency. This constituted the human-factors systems problem—to identify areas of operational slack and bottlenecks in the system, and to develop "quick-fix" recommendations to expedite the placement processes.

Setting System Boundaries

Total system components identifiable in the total range of geriatric senility problems were of course much too extensive to be involved in the limited mechanical system problems that were addressed in the geriatric placement project. Likewise, many subsidiary problems, unless determined to be direct impediments in the mechanics of placement, could not become involved in the immediate analytical problem.[1, 2] Rather, as determined through preliminary investigative depth interviews, the system components directly involved in mechanical problems of geriatric placement are illustrated in Figure 7. These components, as involved in the mechanics of geriatric placement, essentially circumscribed the system boundaries; geographically, as a sampling study, the system was further circumscribed as an 18-county region in East Central Illinois.

Mission Outcome Goals

Operations-research analyses had established that 9,000 geriatric patients resided as chronic patients in state mental hospitals. Approximately 300 chronic patients turned 65 each year, while an additional 3,000 were estimated as admissions candidates each year. Of these, approximately 75 percent were designated for community placement within 18 months. Thus resources of the entire circumscribed placement system required analysis to determine various alternative design fixes that would accomplish specified outcome goals.

Analytical Methods

Various human-factors system methods may be conceivably adapted to such a human-service mission. However, the classical missions analysis approach, as outlined in chapters 1 and 2, may be somewhat limited when addressing preconceived, formal, ongoing or historically entrenched closed systems. Obviously analytical methods are needed that may be most expediently deployed to penetrate the current system. These might involve such approaches as making systematic observations, role-taking through the system, or bracketing the sequences of operations for interrogation of current key personnel in a "critical-incidents" fashion.

Direct observation of current operations as a method may be prohibitively time-consuming, though a critical human-factors eye may indeed uncover unfortunate snags and inefficiencies in the system, such as unwieldy and excessive forms, unnecessary procedures, etc.

The method of role-taking may also be employed in analysis of current systems. This was exemplified by Rosenhan's work[3] where a number of observers had themselves committed to mental hospitals to record the situations to which patients are typically subjected. The role-taking method, however, may also prohibitively prolong the analysis, while perhaps being more costly than warranted.

The method employed in the geriatric placement project was developed adjunctively to include direct observations and structured

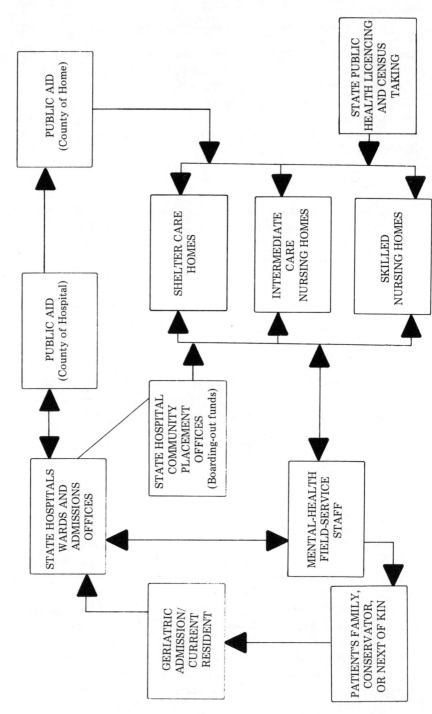

Fig. 7. System Components Involved in Geriatric Placement Mechanics

interviews with key personnel. Through preliminary interviews, detailed flow charting was accomplished using the outline of Figure 7 as a prototype.⁴ Then forms and procedures of the Departments of Mental Health, Public Health and Public Aid were examined, as well as those employed in group-care home processing. Pertinent open-ended interview questions were also developed preparatory to undertaking specific interviews. These were oriented to each critical functional area as follows:

> State Mental Hospital
> > Admissions office
> > Wards
> > Community Placement office
> > Reimbursement and/or Patient
> > Billing office
> > Public Aid Representative
> > Public Health Representative

Mental Health Field Service Staff
Regional Public Aid office
Regional Public Health office
Skilled-Nursing Home operators
Intermediate-Care Home operators
Shelter-Care Home operators

Questionnaire items were designed to elicit both a depth and breadth of information, frequently oriented to a "critical issue"-type approach, e.g., "What do you find to be the most difficult problem in getting your geriatric patients placed?" A total of 14 interviews were completed with 35 interviewees. Of these, 15 were members of the mental health field services staff; two were state mental hospital staff. One was the State Department of Public Health representative, and one the Public Aid representative, both of whom were stationed at the state hospital. The State Director of Community Placement for Mental Health was also interviewed, as well as the regional supervisor for Public Aid and Public Health. A number of extended-care facility operators were also interviewed by class of facility.

Identifying and Documenting Critical Functional Areas

Interviewee responses were recorded by tape or as noted by the interviewer. General information items and problem areas were then drawn from interview data and listed by function. Responses were coded by letters and numerals pertaining to place of interview and item listings for the interview responses. A general outline was then prepared of the overall procedures operating in community placement of geriatric cases. Cases were presented at the state hospital for a "Pre-admissions Examination" (PAE), some of which became admissions, while other geriatric cases were already residents. After admission, critical delays were incurred in the reimbursement/billing procedure. The state hospital, through its Community Placement office, referred a case to be placed or reviewed to the appropriate field service staff (designated by county of origin). The field service staff, however, tended generally to initiate the review.

Cases were required to go through "staffing" at the state hospital and were subsequently referred to the Departments of Public Aid and of Public Health at the hospital. Public Aid determined eligibility for financial assistance or transferred the cases from medical assistance to Old Age Assistance. If ineligible, the cases remained in the hospital or were referred to the Mental Health Department Community Placement office at the hospital for boarding-out money. Public Health assisted in locating appropriate shelter care or nursing homes, but the field staff screened and coordinated cases for specific placement.

From these general procedures, each problem area in placement was then identified and documented by interviewee responses and site of trouble. Reliability was achieved as a function of independent consensus of several different interviewees.

Data were then integrated to provide a systems overview of inter-related problems.

Quick-Fix Recommendations

In identifying the interrelated system problems impeding the overall geriatric placement process, each problem area was then isolated for separate study. The experiences of appropriate staff members were drawn upon, as well as rational consideration of human-factors principles, to formulate design recommendations. Figure 8 presents an overview of resulting system quick-fix recommendations.

Current Openings in Extended Care Facilities. Ample openings in various classes of extended care homes seemed generally to be available in sufficient number throughout the state to accommodate all geriatric cases. Openings, however, were not well coordinated in the placement process. This meant that Public Health should *institute a more prag-matic function as a central clearinghouse, in a timely dispatching of*

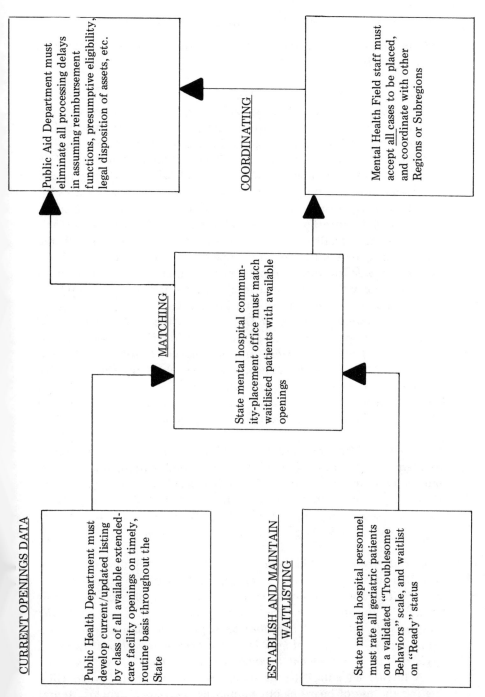

CURRENT OPENINGS DATA

Public Health Department must develop current/updated listing by class of all available extended-care facility openings on timely, routine basis throughout the State

MATCHING

State mental hospital community-placement office must match waitlisted patients with available openings

ESTABLISH AND MAINTAIN WAITLISTING

State mental hospital personnel must rate all geriatric patients on a validated "Troublesome Behaviors" scale, and waitlist on "Ready" status

Public Aid Department must eliminate all processing delays in assuming reimbursement functions, presumptive eligibility, legal disposition of assets, etc.

COORDINATING

Mental Health Field staff must accept all cases to be placed, and coordinate with other Regions or Subregions

Fig. 8. Quick-Fix System Recommendations in Geriatric Placement Mechanics

information to state hospitals on "currently" available openings by class of facility.

Establish and Maintain Waitlisting. A major deficiency at the state mental hospitals appeared to be lack of "match-up" information on patients in terms of placeability. Criteria in screening appeared to be conventional, e.g., nosological, with lack of practical specifications necessary for placement evaluation. *The mental hospital should be required to institute a practical set of mechanics for specifying and waitlisting all geriatric patients when feasible on a "ready status."*

The staffing function in judging the readiness of patients for placement failed to incorporate practical measures of what troublesome behaviors the extended-care facilities would accept. Psychiatrists and other staffing personnel had only superficially observed the patients, and lacked practical intent in placement. Rather, it was recommended that the state-of-the-art in behavioral rating scales be drawn upon to develop a practical instrument for this purpose.[5]

The Community Placement Office at the Mental Hospital should be required to match cases to be placed in coordinating information on current openings available in extended-care facilities throughout the state. Criteria should include, when desirable, placement as close as possible to county of origin when family ties were important; however, such consideration should be of only secondary determination when otherwise impeding the placement process.

Coordinating Placement. Mental Health Field Service staffs often failed to place referred cases due to insufficient information, financial entanglements, or when patients' families lived out of their region. *Field staffs should therefore be required to establish arrangements when necessary to coordinate with families and agencies as necessary to place patients who originated out of their region.*

Financial Assurance. Placement was often impeded due to lack of boarding-out money, delays from the reimbursement claims coordinated with Public Aid, estates and assets being complicated with joint ownership, the snarling of eligibility for Public Aid, etc. *The Public Aid Department should therefore be required to eliminate all processing delays in establishing firmly and expediently eligibility of all geriatric cases for payment to extended-care homes.*

A further advantage to the state was also highlighted when eliminating the use of boarding-out monies by the Department of Mental Health. Restricting financing to the Department of Public Aid meant that half of such costs would be recovered from federal financing.

Coordinating Implementation

Quick-fix recommendations for such human-service systems must, of course, be coordinated at appropriate administrative levels to assure due consideration of recommended fixes for improved system operation. The systems design problem, as previously discussed, is no less formidable within the scope of such operations than it is in broader-based systems. Where single-point authorities are lacking (as they generally are even out of the governor's office at a state level), a functional format may be indicated about which various administrators might be required to react. This might include a statement of each quick-fix recommendation, its supportive rationale and/or documentation, and a required statement of action to be taken by the appropriate administrator with reasons for such action. Such required responses on the part of administrators might avoid personal irrelevancies, while retaining the necessary systems perspective for design fixes.[6]

Follow-up Evaluation

The extent to which a human factors follow-up evaluation might be completed is, of course, dependent upon the extent of cognizance one has over the final outcomes. This might simply involve the monitoring of key indicators over time—in this case the level of geriatric population in state hospitals. With more extensive involvement, more detailed monitoring of outcome for each design recommendation may be possible, while redesign or refixing may become necessary.

GENERAL HUMAN-FACTORS APPLICATIONS TO HUMAN-SERVICE SYSTEMS

If such a human-factors contribution to human-service systems design is to be taken seriously, it must continue to prove its ultimate worth in functional outcome as well as dollars and cents. The geriatric placement study indeed revealed multiple areas of deficiency. Quick-fix recommendations resulted in major cost saving to the state through recovery of Public Aid monies from the federal government. Troublesome behaviors retarding the flow processes in geriatric placement were clarified and developments initiated for improved screening practices. The multiple interactive elements impeding the process were also brought to the fore, thus further enhancing the placement process.

However, wide acceptance and even recognition of the human-factors profession in design of human services are only slowly developing. Hoos[7] has critically assessed systems design accomplishments, and found only weak arguments to support use of systems methodology. Indeed, system design accomplishments cannot be pointed to as glowing achievements in cost-management areas. Overruns may be evident on almost all ventures of the Department of Defense where systems methods were first espoused and practiced. In such criticism, however, the fact of material accomplishment cannot be overlooked in the fantastically complex weapon systems, in manned lunar landings, orbital laboratories, space shuttles, etc. Excessive expenditures, on the other hand, might also be attributed to too little, rather than too much, use of appropriate expertise.[8]

While human-service areas have yet to benefit in large measure from the systems approach, the technology is available, and is but to be summoned to prove its worth in human-service missions.

NOTES TO CHAPTER 4

1. Such subsidiary geriatric problems may include preparing the patients for transfer, their reactions to institutionalization, mortality rates at transfer, etc. Though of fundamental import in the total geriatric system, expeditious processing in placement mechanics would not necessarily involve such considerations.

2. M. Manson and C. Engquist, "Adjustment of Eighty Discharged Geriatric-Psychiatric Patients," *American Journal of Psychiatry* 117 (1960) :319–22; L. Novick, "Easing the Stress of Moving Day," *Journal of the American Hospital Association* 41 (August 1967) :64–74; M. Lieberman et al., "Psychological Effects of Institution-alization," *Journal of Gerontology* 23 (July 1968) :343–53; J. Brudno, "Experi-mental Approach to Services for the Ready-to-Admit Applicant to a Geriatric Home and Hospital," *Journal of the American Geriatrics Society* 16 (May 1968) :597–602; B. Stotsky and J. Dominick, "Mental Patients in Nursing Homes. I. Social Depriva-tion and Regression," *Journal of the American Geriatrics Society* 17 (January 1969) :33–62; M. Linn et al., "A Social Dysfunction Rating Scale," *Journal of Psychiatric Research* 6 (1969) :299–306; F. Frankel and E. Clark, "Mental Health Consultation and Education in Nursing Homes," *Journal of the American Geriatrics Society* 17 (April 1969) :360–65; D. Kay et al., "Mental Illness and Hospital Usage in the Elderly: A Random Sample Followed Up," *Comprehensive Psychiatry* 11 (January 1970) :26–35.

3. D. Rosenhan, "On Being Sane in Insane Places," *Science* 179 (January 1973) : 250–58.

4. Detailed system descriptions of flow sequences, technical and bureaucratic processing in geriatric placement mechanics were developed in a report prepared by the Adolf Meyer Center in Decatur, Illinois, "Geriatric Placement into Extended Care Facilities," Report RER-21, 1971.

5. I. Handy, "Readiness for Psychiatric Hospital Release as a Function of Social Efficiency and Psychopathology," *Dissertation Abstracts* 30, no. 9A (1970); W. Craig, "Scales for Nursing Observation of Behavior Syndromes," *Journal of Clinical Psychology* 26 (January 1970):91–97; C. Ryder et al., "Patient Assessment, an Essential Tool in Placement and Planning of Care," *Health Services and Mental Health Administration Health Reports* 86 (October 1971):923–32; J. Wittenborn, "Wittenborn Psychiatric Rating Scales," in *The Fifth Mental Measurements Yearbook* (Highland Park, N.J.: Sryphon Press, 1959), pp. 210f.; J. Overall and D. Gorham, "The Brief Psychiatric Rating Scale," *Psychological Reports* 10 (1962): 799–812; E. Gruenberg et al., "Identifying Cases of the Social Breakdown Syndrome," in *Evaluating the Effectiveness of Mental Health Services*, Milbank, New York, pp. 150–55; M. Linn and L. Gurel, "Initial Reactions to Nursing Home Placement," *Journal of the American Geriatrics Society* 17 (February 1969):219–23.

6. J. Burgess, "Ego Involvement in the Systems Design Process," *Human Factors Journal* 12 (February 1970):7–12.

7. I. Hoos, "Systems Techniques for Managing Society: A Critique," *Public Administrative Review* 33 (March 1973):157–64.

8. K. Teel, "Is Human Factors Engineering Worth the Investment?" *Human Factors Journal* 13 (February 1971):17–21.

5

Human-Service Subsystems

The foregoing chapter on human-factors systems analysis and quick-fixes in a human-service mission presents perhaps typical isolated instances of possible applied systems management. In order to acquire a detailed perspective on the more total human-service subsystems currently operating in society, one such subsystem component was studied at some length. In this case, the mental health subsystem operating in east-central Illinois was taken as fairly typical and representative of such subsystems.

A MENTAL HEALTH SUBSYSTEM

Mental health subsystems have evolved through historical custom and practical response to deviant and disturbed behavior. Traditionally, the community reaction to mental problems, as with criminal, has been to extrude the offender. Figure 9 illustrates such a subsystem where cases enter from any of a number of the subsystem components. The sheriff or police may pick up "mentals" from an episode of civil or domestic disturbance or complaint; or a petition may be filed with the state's attorney, or the case may be referred as mentally ill in a child-abuse case from the Department of Children and Family Service (CFS). Patients may voluntarily present themselves at the Mental Health (MH) clinic or state hospitals, or be referred by the Department of Public Aid (DPA), or private physicians to general or private hospitals. They may then be referred to the Adolf Meyer Zone Center (AMZC), an inpatient resident unit. In police or sheriff referral, they may be incarcerated in the jail until a sanity hearing is held and certificated as "in need of mental treatment" by a physician. They may then be transported to a state hospital by the sheriff's staff.

With the advent of the Community Mental Health Act of 1963, an increasing involvement of other agencies, such as the Division of Vocational Rehabilitation (DVR), began to operate in the mental

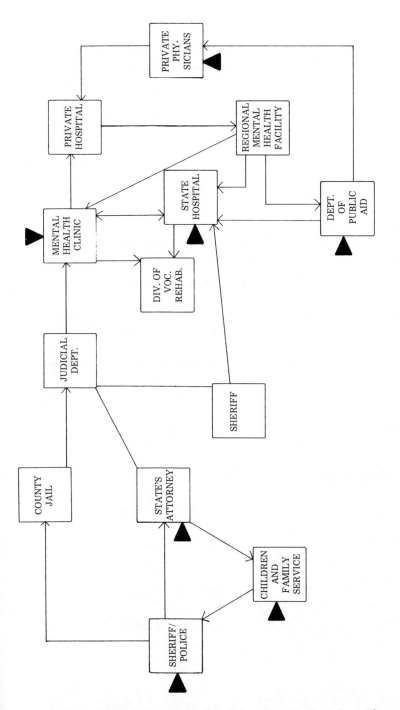

Fig. 9. Typical Mental Health Subsystem, Current Case Flow

▲ Entering from Outside Subsystem

health subsystem. Thus, a community mental health subsystem development was predicated on the necessary interaction of virtually all health, education and welfare social agencies. However, this multiplicity of agencies has merely served to confound the service delivery operation; each agency exercises jurisdictional responsibility and control over decisions and actions which profoundly affect community mental health; yet each acts independently as a discrete functional service, which may or may not find a case eligible to be served. The net result becomes a disjointed, fragmented delivery system, with multiple failures in providing continuity, and comprehensiveness of appropriate services. This may be illustrated from a study of coded department processes. The following prescribed charters, restricting the respective agency's definitive functions, were determined through structured interviews of agencies in east-central Illinois:

Mental Health Clinic—treat emotional problems, psychoses and neuroses.

Children and Family Service—rehabilitate or prosecute parents and arrange for care of their abused, neglected and dependent children.

Circuit Judge—preside in court and rule on felonies and major civil cases.

Associate Circuit Judge—rule on juvenile delinquency and mentals.

Magistrate Judge—rule on misdemeanors and minor civil cases.

State's Attorney—prepare evidence for prosecution of persons involved in criminal acts.

Probation Officer—investgate petitions for probation and monitor probationer.

Division of Vocational Rehabilitation—screen eligibility, coordinate and administer rehabilitation programs for the physically and mentally handicapped over 16 years of age.

State Employment Service—refer qualified applicants to employers.

Unemployment Compensation—process unemployment insurance claims.

Social Security Administration—process retirement and disability insurance claims.

Veterans Commission—assist veterans and survivors to secure entitlement benefits.

Public Aid—determine, process, and monitor eligibility for financial aid.

Housing Authority—provide housing for low-income families.

Public Health—prevent and control widespread diseases.

Office of Economic Opportunity—encourage low and marginal income groups to improve standards.

Prison—deter offenders and remove them from society by isolation and confinement.

In each case, the agency's charter, its facilities, the personnel skills on its staff, its expectancies and habitual practices and routines tend markedly to limit the input of clientele and service effectiveness.

A community mental health service subsystem, as well as other health and welfare service subsystems as described in chapter 2, can operate with only minimal effectiveness in the face of such restrictive charters. An individual discharged from a state hospital may require living accommodations, and often at least interim financial aid, a job or job training, family and social support, behavioral and emotional therapy, and a number of such service provisions for stable maintenance in the community. Such essential service requirements obviously cut across multiple service agency boundaries ranging from Public Aid, Division of Vocational Rehabilitation, and employment services to family counseling services and the Mental Health Clinic.

The health-and-welfare services of an urban county in east-central Illinois were studied in some depth to determine the interrelationships among the agencies in providing comprehensive services to clients in need. The agencies, studied through in-depth interviews with staff supervisors and from file searches, are listed below.

Division of Vocational Rehabilitation (DVR)
Mental Health Clinic (MHC)
Adolf Meyer Zone Center (AMZC)
Department of Public Health (DPH)
Department of Children and Family Service (CFS)
Department of Public Aid (DPA)
Township Relief (TWNS SUP)
Illinois State Employment Service (ISES)
Unemployment Compensation (UNEMP)
State's Attorney (STATTY)
Court (CRT)
Sheriff (SHER)
Probation Department (PROB)
Community Medical Clinic (COM CLIN)
Visiting Nurses Association (VNA)
Office of Economic Opportunity (OEO)
General Private Hospitals (PVT HOSP)
State Hospitals (ST HOSP)

Public Education (PUB ED)
Veterans Commission (VET COM)
Social Security Administration (SSA)
Family Services (FAM SERV)
Salvation Army (SAL ARM)
Crippled Children's Society (CRIP CHILD)

It can be empirically demonstrated that so-called "mental illness" is often more an embedded poverty function than a real illness.[1] The origins of mental illness can therefore be found to reside predominantly in the sociocultural environment which aggravates or perpetuates the condition, in failure to deliver timely, appropriate and effective services. It can further be seen that the boundaries of a community mental health subsystem must be permeable, and interface extensively with the more total health-and-welfare service system. Yet the matrix of Figure 10, showing typical referral patterns, tends to indicate discrete if often not rigid boundaries among the component subsystems. The mental health clinic, Adolf Meyer Center, state hospitals, and general hospitals' psychiatric wards make up the essential components of the mental health subsystem. These can be seen, from inspection of the matrix, to refer only in a fragmented way to financial, vocational training and other service functions. Likewise, the law enforcement and criminal justice service components are subsystem bound, as are the vocational training, financial support, employment, and public education subsystems. The substantive operations of these individual agencies also reveal this in the constraints of their operations. When agency service staff were questioned as to their referral functions, typical reactions were as follows:

"We do a lot of referring within the Department."
"OEO people are not qualified to provide social services. The Adolf Meyer Center is also of no help as an agency—just a lot of make-work committees."
"We have people we don't know what to do with."
"There are no referrals after probation."
"A staff member may refer a needy person to Township Relief or some place out of the goodness of his heart."
"Referrals are made only to the State's Attorney, Sheriff's Department and Police. There is no agency interaction with this office." (Circuit Judge)

DYSFUNCTIONAL SUBSYSTEM STRATEGIES

The consequences of these frequently rigid subsystem boundaries

Numbers Referred to Other Agencies

AGENCY	No Request Service	No. Not Qualified	No. Provided	Total Referred	DVR	M.H. CLIN	AM/ZC	DPH	CFS	DPA	TWNS SUP	ISES	UN EMP	ST ATTY	CRT	SHER	PROB	CATH CHAR	COM CLIN	VNA	OEO	PVT HOSP	ST HOSP	PUB EDUC	VET COMM	SOC SEC	FAM SERV	SALV ARMY	CRIP CHILD
Div. of Voc. Rehab.	791	654	137	416		30			20	16	16	354									40	100	35						
Men. Health Clinic	519	78	441	195	5																		14						
Adolf Meyer Zone Center	340	85	255	160	16	108				8	30																		91
Dept. Pub. Health	16320	12566	3754	763		2			2	2	6			3					3	645				2					
Child. Fam. Service	218	0	218	121		15								3															
Dept. Publ. Aid	1490	745	745	1010	123	15			10	379				3		29	41	50	40	140	68				75	24			
Township Relief	1500	1200	300	812	16						300					3		314							100	30	150		
Ill. State Emp. Serv.	11802	5901	5901	102	16									3										11	75				
Unemploy. Comp.	2010	201	1809	1416								1416																	
State's Atty	28007	4007	24000	16726		10				8		2			2793	13293	505	115				60					15		
Court	18618	8192	10426	1767		10										1477	200												
Sheriff	29611	24873	4738	12598		20			20	2	15			4219	8192			115											
Probation	831	0	831	80												80													
Catholic Charities	2300	0	199	150					10											140									
Comm. Clinic	4000	0	4000	128				3	15											30						80			
Visit. Nurs. Assoc	7050	0	7050	9				9																					
Office of Econ. Oppor.	1376	798	578	1309	15				5	826	274	1						1		1					1		200		
Private Hospitals	28800	0	28800	3002	16		173	1905		500	405																		
State Hospitals	364	0	364	103	16		27			15											54								
Public Education	21600	0	21600	2316	48			2000	20	15							85												38
Veteran's Comm.	2500	2250	250	394							30	354	10																
Social Security	1500	120	1380	82	23					15	30		10														4		
Family Serv.	444	0	444	160					20											140						16			
Salvation Army	1380	0	1380	0																									
Crip. Child	374	0	374	0																									

*Data obtained from local files, interviews, and estimates of local agency supervisors, 1967-1968

Fig. 10. Yearly Health-and-Welfare Service and Referral Matrix

and fractured service functions are often in creating or confounding an adverse life situation for the client. Within the confines of the community mental health subsystem, for example, many administrators tend to define development goals primarily in terms of reducing the number of extrusions from the community to state hospitals. This also means moving patients residing in state hospitals back into the community. One consequence of such a monolithic goal structure is the arranging of incentive funding for state-aided clinics to maintain patients in the community, since grant funds are reduced if patients are committed to state hospitals from the catchment area of the clinic. Based on such singleness of purpose, one episode was revealed in which a patient was maintained in a hotel room where he secluded himself while his medication was passed under the door. Another perhaps typical circumstance in placing patients in the comunity requires that the Public Aid Department assume their financial support. Maximum assets cannot exceed $500 to be eligible for public support. Consequently, the "Reimbursement Section" of the Department of Mental Health is charged with suing the patient, his conservator or next of kin, to reduce his assets to a level of eligibility for Public Aid and community placement. In the process, the mental health service, presumed to be an ameliorative function, becomes a harassment generating further mental health problems.

A multiplicity of such situations may be identified upon further study as a consequence of the "locked-in" boundaries of human-service subsystems. The amount of Public Aid payments per patient to group care or nursing homes is based on level of care needed in bedridden service, feeding, tending body functions, etc. The incentive is thus to maintain a high level of care with a converse low level of patient functioning. The very act of diagnostic labeling in mental health may, in fact, induce such an adverse reaction as a function of subsystem boundaries.[2]

Rigid subsystem boundaries may, no less than Warner[3] has decried, deny the rights of full citizenship to the poor and the mentally ill. Charity and welfare services are chronically undermanaged, begrudgingly administered, and raked with incompetence, abuse, fraud, political expediency, and a confused urgency for reform. Obviously the desirable design of a community mental health subsystem would include an open health and welfare system within which mental health service functions could be carried out with exactitude and proficiency.

A REQUIREMENTS VERSUS MEANS APPROACH

The present mental health subsystem is thus traditionally status-bound and means-oriented, and has virtually *never* been subjected to a design analysis. Such traditional professions as psychiatry, social work, and psychology have dominated the field, while the operational institutions of mental hospitals, clinics, and private medical practitioners largely determine objectives. Objectives when treated may be seen to be implicitly derived from predetermined means, e.g., prevention and treatment of "mental illness." The formulation of social goals, developmental, restorative, rehabilitative or therapeutic objectives, with derived service requirements is either neglected or such considerations. are obscured by the overwhelming precedents of tradition. Professional pressures, for example, are necessarily inherent in the traditional medical model and the associated team of treatment specialists in mental health. Indeed, the force of traditional pathways may be responsible for failure to establish progressive and clear-cut policies from which common developmental goals may be derived. The classical treatment emphasis in mental health is traditionally clinic- and hospital-centered, while current emphasis is shifting to social intervention and community responsibility. Cross-agency networks and federated agency functions therefore become of central importance to the community mental health movement. Yet, the means constraints imposed on mental health developments becomes formidable barriers to effective design and operation. Thus, an approach is called for that is based on requirements derived from objectives. Efficient administrative development of the complex interacting networks inherent in a community mental health program may, in fact, be possible only when such technology is employed.

A human-factors systems approach at the outset must involve strictly a statement of requirements. When a design problem is approached with given means solutions in perspective, it must indeed be designated a "means approach." Only when various alternative means are to be considered in solution, from which a more optimal one may be selected, is a systems or requirements approach involved. Requirements analysis becomes the process of determining successively lower levels of requirements and constraints for which subsequent more detailed design solutions can be achieved: Design is directed toward acceptable real solutions to given sets of requirements and constraints. The process then continues through different levels to the point where the design is completed.

Delineation of basic requirements for a system, as discussed in chapter 2, presupposes an iterated statement of policy. From a general policy is developed the preliminary criteria for design. Then, proceeding within this analytical framework, a requirements analysis may be completed. The analysis consists of descriptions of "input states," and a specification of desired "output states." System requirements are identified as derivatives necessary to accomplish the output state (see Figure 11). From specified requirements, progressively refined details of systems design become clarified at each step in analysis. The criteria bearing on each design decision in selecting an alternative to a system facility, process, or operational skill are those of cost, reliability, and effectiveness.

Policy

The operating policy must be either implicit or explicitly stated to provide the fundamental scheme upon which design parameters are to be based. In national defense, for example, the policy of Switzerland is one of nonaggression. Within the Swiss arsenal are none but defensive weapons, e.g., no bombers. In the United States, the policy in the cold war has been one of retaliation, requiring weapons of the shortest possible reaction time after early warning.

In mental health, a policy may be proposed for hypothetical development purposes as follows:

> "The incidence and prevalence of mental disorders are probabilistic events occurring as functions of multiple social variables. Such disorders must be seen as a conglomerate function of multiple societal needs of population subgroups. Boundaries of a mental health subsystem are thus found in the total sociocultural environment, in which are generated the requirements for effective delivery of health-and-welfare services. Services must systematically relate to population subgroups, and serve to eliminate or reduce the incidence and prevalence of these designated disorders."

Requirements Analysis

In accomplishing a requirements analysis from the policy statement, basic analytical objectives may involve the specification of changes to be performed on each input state. Measurable criteria of an output state must be specified, together with the limitations or constraints imposed on these. The analyses may proceed on the basis of the framework illustrated in Figure 12 as follows:

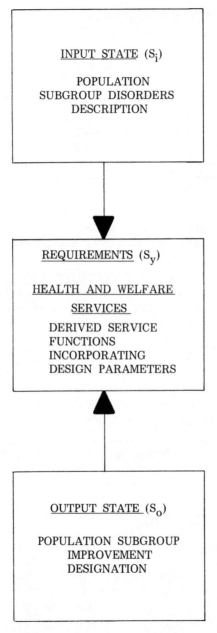

Fig. 11. System Requirements Derivation (S_y) from Input-to-Output Change State

a. Identify objects (or population) upon which the system is to act. (S_i) Subgroups are identified, which may be generically defined as "high risk," "impaired," etc., groups to be served.

b. Identify relevant input state variables, e.g. family grouping, income, living accommodations, etc., as pertinent to the output goals. Living situations, family conflict, etc., would be important considerations, for example, in relocating mental patients in the community.

c. Specify state change requirements as sufficient to identify definitive services necessary to meet specific needs, e.g., a skill to be acquired, a job to be found, a home to be located, etc. (S_o)

d. Establish criterion measures of system performance for determining adequacy of state change effects, e.g., tenure measures for independent living, social and recreational participation, reduced incidence of impairment, etc.

e. Identify constraints, e.g., reorganization costs, vested interests, etc., which must be considered in design (S_y) .

f. Rank order variables according to their system importance, e.g., service to the highest risk categories of population within a given target community.

g. Rank order constraints according to their system importance and formidability, e.g., community prejudices, inertia of needy groups, etc.

h. Identify problem areas, e.g., from f through g above. Set aside for special studies with specific study objectives indicated.

i. Document analysis for guidance in the development sequence.

Preliminary Criteria

From such a requirements analysis, preliminary criteria may be designated upon which development of system characteristics might be based. A mental health subsystem, thus, must be so configured as to provide essential operational effectiveness in service to designated socioeconomic subgroups. Design characteristics would include, but not necessarily be limited to, such system parameters as follows:

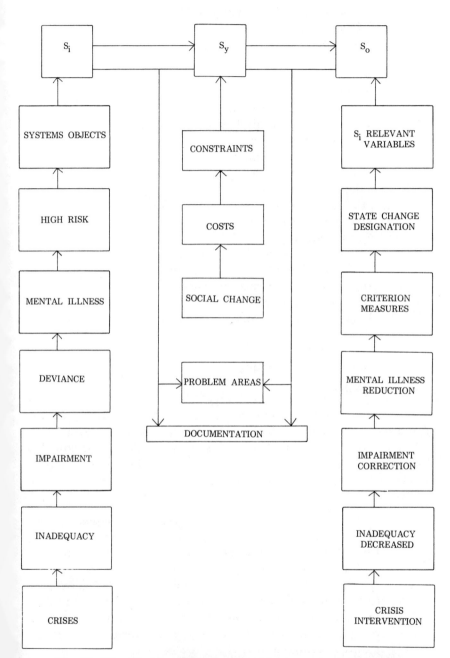

Fig. 12. The Health-and-Welfare Requirements Analysis Process

a. Accessibility and acceptability by client
b. Timeliness
c. Reachability
d. Reliability
e. Financial assurance
f. Minimal costs
g. Citizen responsibility and control
h. Continuity and control of referral procedure
i. Administrative unity
j. Acceptability by community
k. Emergency provisions
l. Safety and protection
m. High-risk subgroup intervention efficiency
n. Optimal employment of manpower/skills
o. Queue efficiency
p. Optimal processing time/minimal lags
q. Optimal suitability/effectiveness of service
r. Information Processing efficiency
s. Decision-making efficiency
t. Remission stability and permanency
u. Panoplied services for coherent coverage
v. Adaptability
w. Renewal and updating

Functions Analysis[4]

When the subsystem requirements are established, and grouped on a logical and practical basis into subsystem components, a functions analysis is then completed. This is a technique whereby the abstract subsystems are analytically divided into smaller manageable performance elements. Steps in functions analysis may be outlined as follows:

a. Delineate performance at each level of development, e.g., detect need, screen, direct individual to help agency, etc.
b. Identify all essential variables within a performance level, e.g., cues, indicants and facts necessary in a referral process.
c. Specify interrelationships and sequences among subsystems (contingencies, dependencies, alternatives, etc.) , e.g., in referral to another service function, the reliability with which the patient contacts the agency, the usefulness of the service compared with another, etc.
d. Establish boundaries of each function, e.g., a referral activity, record-keeping, a therapy element, etc.

   e. Specify constraints, e.g., facilities required, costs, available skills, special equipment and training required, etc.

   f. Prepare functions flow diagrams.

     (1) Indicated continuity. Account for all gaps.

     (2) Provide an overall view relating each subdivision.

     (3) Delineate significance to required output rather than merely means.

     (4) Examine for optimal points of integration of data networks, administrative unity, etc., between subsystems.

     (5) Examine functional requirements for skill levels and number of personnel.

     (6) Examine for training implications.

     (7) Show logical continuity.

     (8) Narrate details.

The functions design activity becomes essentially a series of design-decision points successively converging on physical means. Free play is given in consideration of possible design alternatives, with consequences examined at each level of detail in terms of the contribution each function makes relative to its cost. During the functions analysis phase, the preliminary design criteria are applied, such as acceptability to a patient when reception is made by operator or by a mechanical device, the efficiency of queue or waiting lines, etc.

Each function in effect structures the design problem, with design decisions made for each function before further subsystem development proceeds. Design proceeds on the basis of a "detailing-means-integrating" process. Detailing is the process of delineating lower elements of performance from a next higher level. Means allocation is the assignment of an actual physical operation to accomplish the element of performance; integration then becomes the process of synthesizing or bringing together all elements of performance to form an operational whole.

The next lower level of analysis then addresses variables of the selected means. If decisions cannot be made between alternative means for a given level of analysis, each alternative may be carried to a next lower level, and projected for consequence details in trade-off analysis. A matrix of functions may be used for embracing the problem as illustrated in Figure 13. Trade-off criteria may then be applied to the design, relating to development time, costs, quantity of service, queuing, staff training required, stability or variance of output state, etc.

Functions

		Intake	Diagnosis	Therapy	. . .	Follow-up
	Nurse	PX	P	PC		PC
	Psychiatrist	P*C	PC	PC		C
	Social worker	PC	PC	PC		C
	Clerk	X**	X	X		P
Means	.					
	:					
	Equipment/ machine	C***	PX	PX		PX

*P   = Role in state-change process
**X = Role in controlling the process
***C = Alternate role

Fig. 13. Functions-Means Allocation Matrix

Performance Specification Analysis

The next step, following a functions analysis, is determination of specific levels of performance required. Sequences in the intermediate state-change processes are first specified. A patient's processing may involve initial contact and filing of information, screening his application, the administration of individual and social therapy, his progress evaluation, after-care services and continued follow-up. Performance analysis thus involves the following:

a. Determine specific state variables and parameters needing change, e.g., abusive or other unacceptable behaviors, dependency, etc.
b. Specify criterion measures, e.g., ratings of improved behavior or community adjustment, sociometric scores, etc.
c. Identify essential sequential state changes required, e.g., job and social skills acquisition, interim financial support, grooming, employer acceptance, etc.
d. Delineate supplemental or parallel services, e.g., recourse to training if job placement fails, etc.
e. Establish parameters of state change requirements.
   —accuracy, e.g., effects on general adjustment criteria, etc.
   —performance time, e.g., duration of service, etc.
   —frequency, e.g., number of sessions, etc.

—location, e.g., in home or community setting, in hospital, clinic, etc.

f. Develop descriptions of means to accomplish required performance according to specifications. For synthesis purposes, a means matrix, as illustrated in Figure 14, may be prepared.

g. Determine consequence of error in terms of possible injury or damage to the individual or society, e.g., incorrect diagnosis or drug applications, possible homicidal incidents, property damage, etc.

State-change Processes	Performance Means						
	Social Worker	Nurse	Psychiatrist	Clerk	IBM	Drugs	... N
Initial contact	X			X			
Record & file				X	X		
Screen	X			X			
Individual therapy			X			X	
Social therapy	X						
Evaluation			X				
After-care		X					
Follow-up	X					X	
.							
.							
.							
n							

Fig. 14. Performance Specification Means Matrix

Detailed performance specification draws upon documented evidence of effectiveness, or is predicated on mission-oriented research. The use of partial hospitalization procedures, milieu therapy, social-support cohort group formations, behavior modification technique, etc., for example, may be applied with confidence in performance outcome based on documented evidence of effectiveness.[5] Task analysis methods may also be employed in determining specific skill requirements, manpower, workload, and performance effectiveness from a personnel and cost trade-off standpoint.

Trade-off Criteria

Throughout the requirements analysis process, various design alternatives are being considered, and weighed against each other on a relative cost and effectiveness basis. Design decisions are therefore made in favor of one approach over another if the duration of service is less, or if the level of professional service is reduced, if the effects are more lasting, or whenever the balance in total design perspective may indicate a preferred alternative. It was in such a trade-off process that the use of state hospitals began to decline with other alternatives now being sought in the mental health subsystem, i.e., the chronicity-generating or perpetuating features of a state hospital component, as a case in point, are now realized over the course of a century to be prohibitively expensive and ineffective. In fact, even the most simple of alternatives—the general hospital psychiatric ward—is winning in the balance, i.e., though higher in per diem cost, the much reduced duration of stay favors this as an alternative. More systematic trade-off studies through all such alternative service functions in the mental health subsystem will continue to be sorely needed.

SUMMARY

A requirements approach to the design of a mental health subsystem holds promise of improving the means limitations inherent in the traditional services of coded mental health departments and professional team approaches. In the current commitment to community mental health developments, a breach with such tradition may become mandatory in order to promote a systematic design program for administrative management and control. We now realize that the boundaries of mental health cannot be circumscribed by arbitrary professional definitions, and indeed must involve a more inclusive health-and-welfare network within the scope of prevention, intervention, and treatment. Extensive subsystem studies are needed to identify and relate each design parameter to ultimate system goals for effective administrative management, program planning, and executive decision in the allocation of resources. Through application of the method of requirements analysis, a human-factors and social-engineering approach to mental health becomes possible.

Such social engineering developments must of course be coordinated with the multijurisdictional aspects of government and bureaucratic boundaries. A requirements approach permits examination of alternative means to be considered by legislative bodies and executive personnel, with meaningful criteria upon which to base decisions.

The present policy and professional vested interests in the mental health subsystem appear to be highly resistant to developmental interaction beyond the boundaries of the mental health subsystem. A requirements approach may offer a method for systematic surveying of such subsystems, and, in identifying needs and objectives, may more optimally relate these to more effective outcomes than the conventional, means-oriented modes of disjointed services have provided.

NOTES TO CHAPTER 5

1. W. Eicker and J. Burgess, "Community Mental Health as a Health and Welfare Social System" (Symposium: *Systems Technology, Social Policy and Mental Health*, 76th Annual American Psychological Association Convention, San Francisco, 1968) ; J. Burgess, "Labeling and the Mental Health Enigma," *Etc. A Review of General Semantics* 30 (June 1973) :162–69; A. Graziano, "In the Mental Health Industry, Illness Is Our Most Important Product," *Psychology Today*, January 1972, pp. 13ff.

2. J. Burgess, "Labeling and the Mental Health Enigma."

3. S. Warner, *The Urban Wilderness, A History of the American City* (New York: Harper & Row, Publishers, 1972) .

4. Confusion at this level of analysis may occur with what might be termed a "means review." Functions analysis may erroneously be taken to be a review of means rather than the analytical service derivations from established requirements. For example, functions may be derived from a review of ongoing agency activities such as the eligibility determination processes which are necessarily means-oriented and directed toward circumscribed conventional outcomes.

5. K. Gyarfas, "Psychotheraphy vs. Milieu Therapy: A Three Year Follow-up of 700 Acute State Hospital Patients," APA Paper #6, Department of Psychiatry (Urbana, Ill.: University of Illinois, 1962) ; G. Fairweather et al., *Community Life for the Mentally Ill: An Alternative to Institutional Care* (Chicago, Ill.: Aldine Publishing Company, 1969) ; J. Sommer, "Work as a Therapeutic Goal: Union Management Clinical Contributions to a Mental Health Program" (New York: Sidney Hillman Health Center, 1969) ; G. Paul, *Behavior Modification Research: Design and Tactics* (New York: McGraw-Hill Publishing Company, 1969) ; B. Astrachan et al., "Systems Approach to Day Hospitalization," *Archives of General Psychiatry* 22 (June 1970) :550–59.

6

Methods of Analysis for Operating Human-Service Subsystems

Human-service subsystems may currently be seen, almost without exception, to develop virtually in isolation as the result of response to immediate and pressing sociopolitical demands. Heavy auto congestion forces city planners to consider alternative modes of access to the inner city. Uncontrolled growth of urban population density results in a kind of spontaneous generation of high-rise apartments. An alarming increase in the number of middle-class drug abusers is followed by accelerated program developments to control drug traffic and treat the addicts. Numerous such ad hoc development projects may be cited which tend not only to reinforce the "nonsystem" character of current human-service subsystems, but to confound the total complex of problems, and to compound overall human-service costs and inefficiencies.

In general, the nonsystems problem seems to center about three categorical circumstances: (1) the bureaucratic mystique of those who administer and control each segregated service component, (2) the lack of definitive integrated administrative control at state, county and local levels, and (3) the unfathomable, sheer multidimensional complexity of a genuine systems perspective. While each of these circumstances is formidable in itself, developing clear, concise, easily understood system descriptions should do much to alleviate the complexity of the total problem. That is to say that the power of the bureaucracy and the ineptness and political partisanship of elected administrators might be somewhat circumscribed when system implications are made visible and understandable to the appropriate administrators. The adaptation, development, and exercise of human-factors systems methodology is here suggested to provide a key to the improvement of human-service subsystems. In fact, the practical problems of modern human-services planning and administration may increasingly press such methods upon us—to devise approaches and ways to analyze and evaluate the gross deficiencies and prohibitively costly operations of current human-service delivery subsystems. The present chapter attempts to describe methods and approaches to such analyses, several of which have currently been applied with some success.

METHODS OF ANALYSIS

The gist of the human-service analytical problem may be primarily in the lack of visibility of components. Superficial examination of agency staff or discussion with agency supervisors reveals but minimal knowledge, information, or interest in other agency services and operations. Often the individual service agencies or professions are themselves threatened by the implications of such analyses or assessment, even though they form but one facet of the total service system.[1]

Basic to acquiring a visibility or overview of current subsystem service components is the adaptation of methods for data collection, for organizing the data into meaningful systems terms, and in presenting the subsystem operational and evaluative data in ways comprehensible and acceptable at the administrative level addressed.

Studies were completed during the late 1960s and early seventies in east-central Illinois, to determine the character and extent of human-service subsystems operating in typical urban communities. Methodology was adapted from operations-research and human-factors system analysis approaches, and included the following:

> Compilation of data through the use of structured interviews with key operator personnel (including critical incident techniques).
> Development of analytic models, manual and computer, simplex and complex.
> Orientation to the involvement of appropriate subsystem administrative levels.

Data Compilation

From preliminary examination of a subsystem network of human-service agencies, it became evident that data collected within each agency was highly restrictive. Each agency typically collected only basic eligibility, accounting, and budgetary information required by law. In order to obtain across-agency subsystem data, an entirely new data format was required—one not normally employed by any single human-service agency. Moreover, the information required, for the most part, was not accessible through the usual agency channels or their customary practices in routine record-keeping. For example, source-of-referral information was simply not captured by local offices of the Department of Public Aid. The Department was thus requested to maintain such records for a one-month sampling period in order to obtain these necessary across-agency data.

During the course of interrogating subsystem agency components in a typical metropolitan area, a systematic checklist of items for query was developed as presented in Table 4. The content was developed on an exploratory basis to provide ad hoc data for making the complete subsystem visible. Other information system requirements have since evolved in the course of building meaningful model representations.

Model Construction

Analysis of current subsystem operations requires that compiled data be arranged in some meaningful analytical format that will provide easy visibility for administrative control. Such a process is known as model construction.[2]

A model is a representation of objects, events, processes, complete systems or subsystems, and is employed as a means for envisioning complex circumstances, their prediction and control. Models may be both descriptive and explanatory. Manipulation of a model may be accomplished to test the impact of changes in one or more components of the model on the entire system. In this way, tests may be completed without actually disturbing events in the real world which the model represents. Models may be classified as follows:

(1) *Iconic*—pictorially or visually representing various aspects of a system (e.g., a photograph or model airplane) ;

(2) *Analogue*—employing one set of properties to represent another which a system possesses (e.g., flow of water through pipes is an analogue of the "flow" of electricity through wires) ;

(3) *Symbolic*—employs symbols to designate properties of a system by means of a language or mathematical equation.

In aerospace systems, scale models and wind tunnels are iconic models used to simulate actual flight conditions. In operations research models, a mathematical description of an activity describes relationships among various elements with sufficient accuracy to predict actual outcomes. Mathematical models may be more or less complex, depending upon the situations they are designed to represent.

TABLE 4
STRUCTURED HUMAN-SERVICE AGENCY INTERVIEW CHECKLIST

Parameter of Agency Component for Query	Substantive Operational Dimensions Probed
Organizational aspects	Public laws, charter or policy requirements; structure, history, changes, parallel services, routine and annual reports.
Input characteristics	Geographic area served; population characteristics and types of problems; case load volume and periodic fluctuations; forms and files maintained; source of referrals; eligibility requirements, numbers rejected, referred or not referred.
Processing operations	Fees if any; processing sequence and time required; judgments and decisions required at reception—while service is in process and following service; standards employed; most difficult judgments and decisions to be made and reason for difficulty.
Information requirements	Information received at reception, oral or written and source; information developed, source and processing mode; how used; to whom accessible and for what reason.
Output or product of service	Termination criteria; referrals made, to whom and reason; typical conference or referral action; information transmitted to whom and why; other formal and informal agency interaction; kind and frequency.
Follow-up action	Nature of follow-up service; evaluation criteria, if any; duration of service effects; repetitive routines and why necessary.
Agency sustenance	Funding sources; administrative authority; budgetary constraints; number and kinds of professional staff and skill requirements; agency location and reason; future plans for growth, merger or reallocation.

The model, rather than the system, may be manipulated in a variety of ways while actual systems are left intact and later adjusted with a minimum of disruption. Model building is limited to representative aspects of a system. The process, though extremely beneficial and valuable, is often excessively time-consuming and costly. Thus, within limits, models provide a tool for enhancing judgment in controlling large and complex systems. Models free the intuition, and permit concentration on subsystem problems of particular interest. Creative managers may then test rigorously the implications of new plans, schemes, and ideas. The symbolic subsystem models developed in the present study consisted of those ranging from simple flow charts and matrices to a more complex set of computer and interpretive representations.

Simplex Models. Data derived from the structured agency interviews were developed into various simplex models. Figure 15 presents a simple flow chart depicting the service sequences for the internal and referral functions of a community medical clinic.

The referral matrix illustrated in Figure 10 (chapter 5) presents a simple model of the yearly volume of referral interactions among an aggregate of health-and-welfare agencies. The model permits study of several interacting aspects of human-service subsystem components. It can be seen first that within the aggregate of 25 agencies serving the public, the referral interaction is spotty and fragmented. The Division of Vocational Rehabilitation, for example, with a total of 791 requests for service by phone or walk-in, turned away 654 as not qualified for service by charter, viz., clients were not physically or mentally disabled in meeting the eligibility criteria. Four hundred and sixteen were referred to other agencies, while 238, or more than 25 percent, were provided with no service whatsoever, either directly or in referral. In fact, it was learned from agency supervisors in general that referral functions in these instances were neither routine nor necessarily expected functions of the agencies.

The matrix also points up other aspects of service deficiency. The first nine rows, Division of Vocational Rehabilitation through Unemployment Compensation, includes functional areas of rehabilitation, training, mental health, public health, neglected children, general welfare, and employment. Total number of requests for service at these nine agencies was 34,990. Over half, or 61 percent, were not qualified for service. Only 39 percent could be given direct service, while only 14 percent were referred to other agencies, including both those determined ineligible as well as those given direct service.

A further characteristic shown in the matrix is indicated, in that

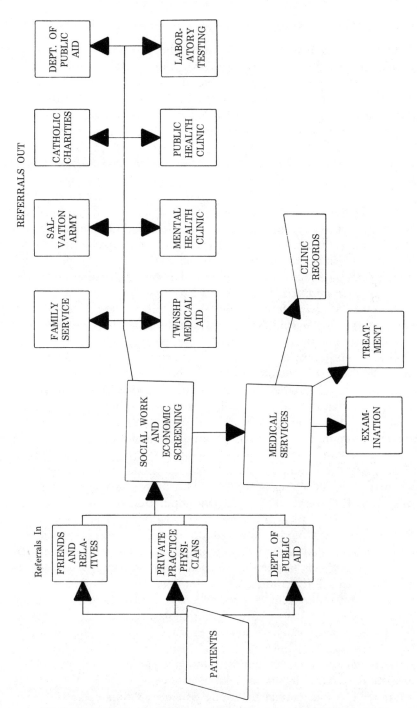

Fig. 15. Community Medical Clinic, Functional Flow Chart

of 23 social problems identified within the community, 17 were in-volved in from two to eight agencies. Cases of illegitimate children were served by the Children and Family Service, the Department of Public Health, the Department of Public Aid, Catholic Charities, and the Continuing Education Center. Alcoholism was served by the Mental Health Clinic, general hospitals, the Meyer Center, the Sheriff, the Salvation Army, and Alcoholics Anonymous. Physical disabilities were handled by the Division of Vocational Rehabilitation, Crippled Children Society, the Illinois State Employment Service, the Department of Public Aid, Social Security, and Special Education. Likewise, emotional problems were served by eight different agencies, including two different state resident facilities. The model shows a peculiar mix of sundry human-service functions that intervene in apparently discrete problems at random arrival times, without con-vergence and correlation of action, which results in high costs and indeterminate measures of service effectiveness.

Complex Models. More complex models incorporate increased mul-tiples of variable dynamics and characteristics of the system or sub-system. Each model is best designed for employment at designated administrative levels, or for specific engineering purposes. These are also called heuristic (exploratory) or stochastic (point-by-point vari-able) models. A number of such complex models were constructed to represent various aspects of human-service subsystems.

A Stochastic Gaming Model. A dynamic gaming model was de-signed to familiarize all levels of administration with dynamics of a human-service system. The game player interfaced at the console of an IBM 1130 computer to administer annual funding, in some proportion, to an aggregate of 30 health-and-welfare agency services for a simulated community of 125,000 population; the agencies served a total of 32 categorical problems of empirically derived frequencies. The programmed model consisted of a series of transitional-proba-bility matrices that provided point-by-point variables of agency ser-vices with problem frequencies. Truancy, for example, was made to increase with adult criminal convictions, as was Aid to Dependent Children; unemployment was cross-correlated with high school drop-outs, crime, neglected children, emotional disorders, etc. In reallo-cating funds among the various agencies through gaming strategies, e.g., decreasing frequency of mental health problems by increasing employment services, the administrator–game-player could observe the increase or decrease in the total social problems complex. A 15-minute computer run was equivalent to one year of real-time operation. Such a model permitted an administrator to see the overall impact on a profile of social problems through various human-service funding configurations. Thus, it purported to sensitize him to the system implications of the human-service functions.

A Human-Services Engineering Model. A discrete-events simulation model for human-service systems and subsystems was developed for use as an engineering tool in human-services design studies. The model was designed by adapting the PL-1 Programming Language for encoding human-service agency operations on an IBM 370 model computer.[3] The model presented referral sources, queuing, internal-processing, and referral-to-external resources, together with costs and processing times. Use of a large computer permitted the running of such detailed representations for several years in a matter of minutes. As a microcosmic model, the designer could vary any of a number of parameters to observe operations and determine potential operating costs. For example, state hospital commitments were reassigned to a newly designed (simulated) day-care center at a general hospital, and provided follow-up service by the mental health center. Chronic patients were moved into group-care homes and independent living in conjunction with day-care and mental health center operations. A number of such perturbations or changes were introduced, while the simulation model optimized for acceptable queuing, provided facilities and staff as necessary, and specified costs, while benefits and other parameters were supplied by the designer in augmenting the simulation with design rationale and theoretical benefits.

Human-Service Analytic Models. Infused with valid and meaningful data, a number of complex models may be constructed that will permit analysis and evaluation of various aspects of human-service subsystems or their components. An open-adaptive systems model, for example, was employed to determine the extent to which an agency component contributed to, or participated as a part of, the more total human-service system.[4] In measuring several of the agency's reported parameters over time, such as referral patterns, staff positions, etc., the dynamic-static character of these dimensions in the agency's operations were ascertained, and determination was made if indeed it were open and adaptive as a community change agent, and responsive to the total needs of the community and to the more total needs of its clients. Such an analytic model permits assessing the quality of the service component as an adaptive and contributing component of the more total human-service system.

Another analytic model developed in the study, termed a "network analysis," employed a data base consisting of client flow through the entire network of services.[5] Baseline data for the model were required in the form of a calendared history of each client receiving services across the system of agencies. This meant that each agency serving its clients was required to provide a common identity for each client across the network of services indicating the periods during which services were provided, as well as the nature and cost of the services. This thus makes visible the entire service sequence, displaying outcomes along with interactions in the serving agencies. Total system measures, often difficult if not impossible to obtain in a service delivery system, were available in the form of cycles and longest paths. The analysis consisted of plotting the pathways for assessment of the cycles and longest paths. The illustration in Figure 16, for example, shows a three-node cycle (A, B, and C), and a longest path consisting of A, D, E, and F.

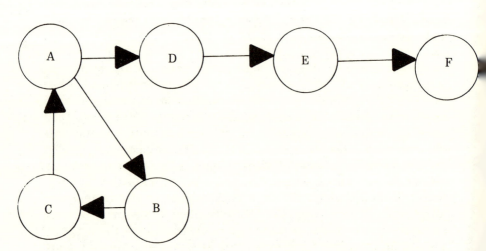

Fig. 16. Illustrative Network Analysis, Patient Flow Model. From John H. Burgess, "Mental Health Service Systems: Approaches to Evaluation," *American Journal of Community Psychology* 2, no. 1:90.

Client service delivery data across autonomous social agencies for the network model were, of course, not available. However, the Meyer Center field staff, serving as advocates for mental patients in obtaining services from a conglomerate of community agencies, maintained patient records of sufficient narrative to construct a service delivery history for each patient. Using such data, service histories for each case were calendared across the multiple service agencies involved with each case, with duration of service identified for each agency. Service network plots were completed, with cycle and longest-path durations calculated for each patient. Statistical measures were then derived from the model for comparison of cycles and longest paths. Results of the calculations indicated that:

(1) More two-node, three-node, and four-node cycles occurred for "unsuccessful" cases, i.e., as judged clinically.
(2) Average cycle length was greater for "unsuccessful" cases.
(3) Longest paths were greater for "successful" cases.

The model thus made visible a more total human-service delivery operation for analysis, and described typical deficiencies in stereotyping services, services gaps, inappropriate services, etc.

Obtaining Administrator Involvement

Human-factors and systems exercises in design and evaluation of human services is of but little avail when extraneous to administrative control and management direction. However, the splintered character of such development control presents a formidable problem. Table 5 summarizes a study of human-service agency aggregates in a perhaps typical urban community of less than 100,000 population. Note the multiple funding sources, and the multijurisdictional administrative authorities operating within the same aggregate of human-service agencies. A perhaps not atypical conflict of such authority is illustrated in the letter of a mayor to the citizens of a small community (see Exhibit A at end of chapter).

Widespread interest within human-service subsystems, however, seems currently to be militating for the consolidation of subsystem functions.[6] Likewise, evaluation has become a matter of increasing concern among subsystem exponents.[7] Yet the subsystem administrative skills, knowledge, education and information continue to be oriented to traditional survival modes. The organizing and operating principles of current administrative authorities are still the traditional ones, dictated by the compulsion to retain stable institutional structures and operations within well-defined boundaries. Rather, a new mode of management is called for—one of an adaptive, rapidly changing, temporary systems character. Task forces need to be organized about human-service problems, drawing from reserve pools of talent for diverse professional and technical skills. Subsystem work groups need to be formed in response to the nature of the problems rather than to programmed role expectations. Management components would become coordinators, serving linkage functions among the task forces. They would continue effectively to utilize the multiplic-

TABLE 5

HUMAN-SERVICE AGENCY AGGREGATES IN A REPRESENTATIVE URBAN COMMUNITY OF 90,000 POPULATION

Human-Service Aggregate and Agency	Funding Source	Administrative Authority
Domiciliary		
Community Development	City tax	City manager, Council, Mayor, & Agency head
Schools	City, county, & federal taxes	School Board and Superintendent
Library	City tax	Mayor & Librarian
Water Dept.	Private pay	Mayor & Agency head
Industry and Commerce		
Community Relations	City tax	Mayor & Agency head
Economic Development	City tax	Mayor & Agency head
Surface Traffic		
Driver License Examiner	State tax	Governor & Agency head
Community Development	City tax	Mayor, City manager, & Agency head
Highway Dept.	Country & federal taxes	Bd. of Supervisors & Agency head
City Engineering Dept.	City tax	Mayor, City manager, & Agency head
Public Transit	City tax & fees	Mayor & Agency head
County Planning	County tax	Bd. of Supv. & Agency head
Air Traffic		
Transportation Dept.	Federal tax	U.S. President & Agency heads
Fire Protection		
Fire Dept.	City tax	Mayor & Agency head
Law Enforcement & Criminal Justice		
Police	City tax	Mayor & Police Chief
Sheriff's Office	County tax	Bd. of Supv., States Attorney, & Sheriff
F.B.I.	Federal tax	President & Bureau Chief
State Police	State tax	Governor & Agency head
State's Attorney	County and state tax	Bd. of Supv. & Agency head
Civil Defense	County tax	Bd. of Supv. & Agency head
Circuit Court	State & city taxes	Governor, Legislature, & Elected Justice
Probate Court	State & city taxes	Circuit Justice
Circuit Clerk	State tax & fees	Circuit Justice
Juvenile & Adult Probation	County & state taxes	Circuit Justice

TABLE 5 (CONT'D.)

Human-Service Aggregate and Agency	Funding Source	Administrative Authority
Public Defender	State tax & charity	Circuit Justice
Corrections Dept.	State tax	Governor & Agency head
Sanitation		
City Engineering	City tax	Mayor, City manager, & Agency head
Garbage Collection	Private pay	County Health Dept. & Agency
County Health Dept.	County, state & federal taxes	Bd. of Supv. & Agency head
Rabies Control	County taxes	County Health Dept. head
Coroner	County taxes	Bd. of Supv.
Health		
Physicians	Private pay	State Health Dept. & individual physicians
General Hospital	Private pay	State Health Dept.
Community Medical Clinic	Charity	United Fund head
Cooperative Health Agency	Federal tax (OEO) & charity	Agency head
Regional Health Council	State	Regional head
Visiting Nurses Ass'n.	Charity	United Fund & Agency administration
Red Cross	Charity	United Fund
Mental Heath		
Mental Health Center	County & state taxes and fees	Bd. of Supv., State Mental Health Region & Agency head
General Hospital Psych. Ward	State & federal taxes & private pay	State Mental Health Dept. and Region administrators
Physicians	State & federal taxes & private pay	Individual physician
Meyer Adult Residence	State & federal taxes	Mental Health Dept. Region Director
Adler Children Center	State & federal taxes	Mental Health Dept. Region Director
Sheriff	County tax	Bd. of Supv. & State's Attorney
State's Attorney	State & county taxes	Bd. of Supv.
Circuit Court	County & state taxes	Elected Justice
Public Welfare		
Public Aid Dept.	State & federal taxes	President, Governor & Agency head
Children & Family Services	State tax	Governor & Agency head

TABLE 5 (CONT'D.)

Human-Service Aggregate and Agency	Funding Source	Administrative Authority
Public Welfare (cont'd)		
Circuit Court	State & city taxes	Elected Justice
Circuit Clerk	State tax & fees	Elected Justice
Labor Dept.	Federal & state taxes	U.S. President & Governor
Vocational Rehabilitation	Federal & state taxes	President, Governor, & Agency head
Veterans Administration	State tax	Governor & Agency head
Veterans Regional Office	Federal tax	President & Agency head
Office of Economic Opportunity	Federal & city taxes	President & Agency head
Human Relations	City tax	Mayor & Agency head
Catholic Charities	Charity	United Fund & Agency head
Salvation Army	Charity	United Fund & Agency administrator
Family Services	Charity	United Fund & Agency head
Township Relief	County tax	Bd. of Supv. & Agency head
Agriculture Dept.	Federal tax	President & Agency head
Social Security Adm.	Federal tax & insurance payments	President & Agency head

ity of pertinent available methods, techniques, and talents in order to orchestrate subsystem missions and development programs.[8]

Presently, however, it would appear that the only recourse to conventional management in the current confounding assortment of bureaucratic administrative authorities and vested interests is in the development of common purposes, goals and objectives. These may largely accrue in response to making operations visible throughout the total subsystem, through the compilation of meaningful baseline data, modeling and operations research, human-factors systems analyses, etc.—in fact, all methods that may be brought to bear on comprehensive system planning needs.

System and subsystem planning is no mere shibboleth, nor does it confront society with the danger of overcontrol.[9] Rather, without deliberation and planning, the freedoms of movement and discourse in human society may necessarily be curtailed. Indeed, without human service analyses and planning, society must expect to be confronted with but one and/or a multiple series of emergencies and crises after another—in traffic, housing, health, energy, and welfare. Real solutions would thus seem to lie in an aggressive and knowledgeable civil management, with a leadership that wisely exploits the best analytical and administrative techniques available for subsystem planning and evaluation.

Strategies and Tactics for Orienting Administrators to Human-Service Systems Design

In orienting the various levels of civil administrators and public management decision makers to systems technology, a number of different methods may be employed.[10] Disseminating systems theory, techniques, and approaches through their communication media may be one such approach—submitting system papers to such periodicals as the *Public Administration Review, Administrative Science Quarterly, Journal of Business Education, The Annals of the American Academy of Political and Social Science, Financial Management, Journal of Comparative Administration, Journal of Human Resources, Quarterly Digest of Urban and Regional Research, Urban Affairs Quarterly, Journal of Public Policy,* etc. Such journals may frequently or occasionally cross the desk of public officials. System articles, if provocative, may excite their interest, if not begin their education in the management and application of system methods. Journal publishers themselves may become instrumental in the systems technology persuasion; if sufficient readership potential is indicated, an especially tailored system journal may be directed to the public administrator, e.g., a journal entitled *Systems Applications in Public Administration.* Government sponsorship of journals such as those now sponsored by the National Institute of Mental Health (*Administration in Mental Health, Evaluation,* etc.) may also be possible. The systems approach may thus be supported through the offices of the Departments of Health, Education and Welfare, Transportation, Justice, the Federal Bureau of Investigation, etc.

Educational institutions may introduce courses on systems engineering in the public sector, as part of a management curriculum in psychology and engineering university departments, seeking federal grants for the explicit development of human-service system study and application.

Another possible tactic is the promotion of system consulting interests to the public sector by the Operations Research Society, the Human Factors Society, the American Psychological Association; indeed, those with systems skills may even be behooved to volunteer their services in advice and counsel if the practical advantages of systems engineering are to be made known.

THE INTRODUCTION OF SYSTEMS SIMULATION AND GAMING TECHNIQUES

In addition to invited speakers on systems approaches applied to public services, a number of simulation and gaming strategies may be introduced. These may be held at regular meetings and conferences of state governors, various public officials such as urban planners, mayors, city managers, and serving personnel including psychiatrists, social workers, nursing groups, community organizers, etc. Special workshops at national, state, regional, and local levels may also be held. Gaming and simulation problems may be designed for each level, or at some mix of levels depending upon the participants. Mixes of public officials may act as players concerned with problems at concrete service levels as well as in general policy formulation. The collective problem solving and deductive reasoning required with abstractions,[11] taken from a human service systems context, can thus produce new insights when working in an atmosphere of differing points of view where compromise solutions may become mandatory.

Various simulation approaches have been discussed earlier in the present chapter. However, simplex and interactive models need not involve computerization. The essential feature of gaming models is that of interpersonal role interactions in providing a lively give-and-take context. At the outset of a game, for example, each player may be required to formulate an explicit set of objectives for his sector of human service operations. Such goals serve as the basis for system development, and, when in opposition to those of other game players, animated discussion may follow. The consequence might, it is hoped, be a more comprehensive systems perspective for the game players.

HUMAN-SERVICE DECISION GAMING

The object of the game is to sensitize the players made up of a mixed group of agency staff personnel and administrators. Different levels of administration may participate, ranging from those in the federal bureaucracy to state governors, regional administrators, mayors, city managers, and various agency heads, e.g., welfare supervisors, police chiefs, psychiatrists, nurses, social work supervisors, etc. A human-service game, as in other analogous situations, is designed for informal interaction. Thus, empathy with other points of view may be developed, while derived solutions may be subject to expert analysis for immediate critique serving both as positive (reinforcing) or negative (change-inducing) feedback to decision-making style and strategy.

Each player may keep his own role as an actual decision maker in the bureaucracy or community, or he may be requested to exchange roles with another in the group. The gaming may then be presented as civil or social problems to be solved. The group may map out a strategy to reduce the rate of onset for mental health, e.g., transferring monies from mental hospitals to community resources with a strategy targeted on occupational training and local hospital care when needed. Such a strategy, with given amounts transferred from the state institutions, is then scrutinized from a systems standpoint and critiqued by systems engineering analysts with the total group; for example, the experts may point out that the simple adaptation process of patients transferred to the community may result in an increased readmission rate to state facilities. The state hospital staffs would be thrown out of work and petition the governor through political pressure to cease such transfer of patients, whom they regard as unsuited for society. Acting-out problems of the patients, and intolerances of vocational training staff as well as potential employers would require major back-up support, increasing costs by 15 to 20 percent, etc. Such constraints would thus serve as "negative" feedback, forcing the game players to regroup and develop new strategies.

Likewise, gaming strategies may be developed to reduce child abuse among the game players made up of judicial officials, child welfare agency supervisors, state and regional administrators (or perhaps a group of governors at a workshop conference). The plan for funding and preventive child care may be formulated by the group, with alternatives to court or agency processing developed, e.g., a parental guidance program on child discipline and varieties of child abuse.[12] The plan, then being subjected to critique by experts in the field and systems analysts, may show a positive gain to be achieved, with reduced court backlog, with redeployment of Children and Family Service case workers reducing job-loss resistances, etc.

Such simulation and gaming approaches may be devised for delinquency, adult corrections, parole and probation, mass transit systems, etc.; gaming system design may be developed for any of a number of such public or human-service systems.

Gaming methods, with and without the use of computer-assisted gaming processes, have been extensively employed in social and environmental simulation. Gamson[13] describes social simulation games (SIMSOC) in training students in social control decision making in the use of police action and political social control. The use of environmental computer-based gaming has been discussed by Philip Patterson of the Environmental Protection Agency, dealing with air, land and water pollutant control.[14] The use of games for systems orientation of students and administrators in urban operations has also been described by House and Patterson.[15]

Strategies for orienting the appropriate players, i.e., the actual decision makers in the human-service delivery process, may be accomplished through the use of such models[16] that were described earlier in the chapter. A basic orientation may best be imparted to human-service providers by informal game playing, with a mix of roles and functional players from all levels of administration. Carefully designed operational models may thus serve both as training instruments in orientation strategies, as well as management tools for program design, implementation and development.

NOTES TO CHAPTER 6

1. Q. Rae-Grant et al., "Mental Health, Social Competence and the War on Poverty," *American Journal of Orthopsychiatry* 36 (July 1966) :652–64; J. Miller, "APA: Psychiatrists Reluctant to Analyze Themselves," *Science* 181 (July 1973) : 246–48.

2. R. Johnson et al., *The Theory and Management of Systems* (New York: McGraw-Hill Book Company, Inc., 1963) .

3. W. Eicker, "Whither or Wither Mental Health" (film, Applied Human Service Systems, Heller Graduate School, Brandeis University, Waltham, Mass., 1971) .

4. R. Nelson and J. Burgess, "An Open Adaptive Systems Analysis of Community Mental Health Services," *Social Psychiatry* 8, no. 4 (1973) :192–97.

5. J. Burgess et al., "Network Analysis as a Method for the Evaluation of Service Delivery Systems," *Community Mental Health Journal* 10 (Fall 1974) :333–44.

6. J. Fishman and J. McCormack, "Mental Health Without Walls: Community Mental Health in the Ghetto," *American Journal of Psychiatry* 126 (April 1970) :

105–10; A. Ivey and J. Hinkle, "A Study in Role Theory: Liaison Between Social Agencies," *Community Mental Health Journal* 6, no. 1 (1970) :63–68; H. Schulberg, "The Mental Hospital in the Era of Human Services," *Hospital & Community Psychiatry* 24 (July 1973) :467–72.

7. P. Levinson, "Evaluation of Social Welfare Programs. Two Research Models," *Welfare in Review*, March 1966, pp. 5–12; A. Levine, "Evaluating Program Effectiveness and Efficiency," *Welfare in Review*, February 1967, pp. 1–11; E. Hargrove, "Program Evaluation: A Program Director's Viewpoint," in *Statistics in Mental Health*, National Institute of Mental Health, Mental Health Statistics Series D, no. 1 (Chevy Chase, Md., 1970) ; A. Wellner et al., "Program Evaluation: A Proposed Model for Mental Health Services," *Mental Hygiene* 54 (October 1970): 530–34; H. Menn, "Developing Principles of Cost Finding for Community Mental Health Centers," *American Journal of Public Health* 61 (August 1971) :1531–35; J. Burgess, "Mental Health Service Systems: Approaches to Evaluation," *American Journal of Community Psychology* 2, no. 1 (1974) :87–93.

8. R. Sorenson, "Manpower System Models in Personnel Allocation Research," *Human Factors* 10 (April 1968) :99–106; W. Knowles et al., "Models, Measures, and Judgments in Systems Design," *Human Factors* 11 (December 1969) :577–90; J. Burgess, "Who Has the Administrative Skills in Mental Health?", *Public Administration Review*, March/April 1974, pp. 164–67; G. Barnhart, "Social Design and Operations Research," *Public Health Reports* 85 (March 1970) :247–50; H. Fox, "Toward an Understanding of Operations Research Concepts," *Management Services*, July 1970, pp. 23–36; C. Martin, "Beyond Bureaucracy," *Community United Way of America*, September 1971, pp. 14–22.

9. I. Hoos, "Systems Techniques for Managing Society: A Critique," *Public Administration Review* 33 (March 1973) :157–64.

10. Sponsorship, of course, remains a major problem. Key public officials must become among the first to assume initiative in promoting such methods. Senator Gaylord Nelson, for example, as have other members of the U. S. Congress, introduced a bill for the adaptation of system methods to the public sector. When such a policy of public management becomes extant at the national level, states, regions, counties, and municipalities may quite readily follow suit. It may therefore most behoove the leadership in systems technology to pursue these influential levels among public officials, eventually perhaps through passage of public laws to tie financial assistance or federal sharing of funds to the proper systems management and design of public services.

11. For discussion of the abstracting process in social gaming, see J. Coleman, "In Defense of Gaming," in *SIMSOC: Simulated Society Participants' Manual with Selected Readings* (New York: The Free Press, 1969) , pp. 27–29.

12. K. Alvy, "Preventing Child Abuse," *American Psychologist* 30 (September 1975) :921–28.

13. See also J. Raser, *Simulation and Society. An Exploration of Scientific Gaming* (Boston: Allyn and Bacon, Inc., 1969) .

14. P. Patterson, "Comments on a Few Computer-based Environmental Gaming Models" (Environmental Studies Division, Environmental Protection Agency, Washington D.C., 1973) .

15. P. House and P. Patterson, eds., *An Environmental Laboratory for the Social Sciences* (Environmental Protection Agency, Washington, D.C.) ; L. Summers

et al., "The Design and Use of Urban Analysis Games in Education," *The Journal of Engineering Education* (in press).

16. W. Knowles et al., "Models, Measurement and Judgments in System Design," *Human Factors Journal* 11 (December 1969) :577–90.

EXHIBIT A*
ILLUSTRATIVE CONFLICT OF
CIVIL ADMINISTRATIVE AUTHORITIES

THE MAYOR SPEAKS. . . .
To the Citizens of Monticello, Illinois:

I'm questioning the stability of our City Council. These men were "elected by the people, for the people"—yet the 6 months I've been Mayor they spend more time "picking me apart," having closed meetings to plan their strategy, and caring less about city problems.

I spoke to our senior Council Member Bill H— at the last meeting about getting down to city business. He agreed but a few days later he wouldn't back me in working out a schedule for the City Police. Are the City Police to tell us what hours they will work? Are their second jobs more important? Did you know that there were times during the nights and day, the city wasn't police protected? I tried to get the police a better wage. At that time again, Mr. H— stated, "if they don't like what they're making let them quit." Yet, the personnel committee other than Mr. G— wouldn't back me in a work schedule. A couple months ago they told me to go ahead with the schedule.

The Council has changed the city ordinances twice since I've been in office. Now they are working on the third time to keep me off committees. So as Mayor I wouldn't have any authority.

There were "certain factions" that wanted to recall the city election at the time I was elected, but found it couldn't be done legally. So they are now trying to provoke me into resigning. As long as I'm doing right for the taxpayers I'll stay in and fight, but I need your support!

Mr. C— has been working on the "liquor code" for approximately eight weeks. He stated he could accomplish this in ten days or two weeks. Why is he holding back? Could it be "someone" is telling him what to do?

* From *the Piatt County Journal—Republican,* November 9, 1973.

I believe in doing my best for the growth of Monticello, if not I wouldn't have cleaned up and built two businesses here.

There are many new people in Monticello; we cannot expect the city to lie dormant!

I believe in presenting the facts to the people, that people who are fair-minded will see the facts and stand behind me.

Speak to your City Councilman. Attend your next Council meeting.

Mayor M.

Part III

Applying System Methods
to Human-Services Design

Human or public service requirements in the United States must be identified for operational projection in advanced planning, as well as for current design improvements of existing systems.

Steeped in tradition as they are, with personal vested interests and professional credentials claiming precedence, privilege and exclusive capability to serve, e.g., medicine in psychiatry, legal expertise in the justice system, social work in welfare and corrections, etc., improvements may best be sought through the existing body politic and pressure groups. Functional alternatives must be openly considered and evaluated by those who themselves are involved in the service process—only then will acceptance of radically new approaches be palatable, and progress in human services become possible.

Such technologies as operations research, cost benefit analysis, specification design control, and systems manpower training methods are readily available system approaches. If properly applied, these could indeed stand to advance the development of both current and future public and domestic service systems.

7

Projecting Human-
Services Subsystem Requirements

To plan or not to plan is indeed a core issue in human-service systems design. In turn, the potential for a human-factors systems contribution, with its unique technological expertise, must revolve about the extent to which policy is explicated, and planning is seen as central to the proper functioning of society.

The future of human societies is often viewed with pessimism, faulting the government's failure in the effective deployment of limited resources.[1] Policies and means solutions are sometimes pressed upon an administration, based on theoretical constraints and outcomes,[2] while many exponents of systems technology continue to urge civil and industrial management to avail themselves of state-of-the-art techniques and analytical methodologies.[3] A monumental theoretical volume was recently published, under the aegis of members of the U.S. Congress, espousing controlled management and planning.[4] Such prestigious proponents of the broad band of systems technology indeed lend much encouragement and promise to a future of planning. Yet, the second-term Nixon administration declined to sponsor a mid-decade sampling census, which continues to harbinger ill for accuracy in planning in the decade of the 1970s.

Basic planning data are, of course, those of population trends. Projections were badly overestimated for population growth after the first two years of the 1970 census. A near zero growth occurred during the early seventies, with a projected fertility of 2.0 children per woman. Table 6 presents estimated population projections for the year 2000 based on 1973 trends.[5] Hard current census data are, of course, required to obtain accurate stochastic projections. However, the estimates as presented appear to be reasonably likely outcomes.

Goals and opinions in human-services planning may frequently be subject to individual bias and provide only a precarious basis from which to project human-service subsystem requirements. More

TABLE 6
ESTIMATED POPULATION, PROJECTIONS FOR THE YEAR 2000

Projected Parameter	1973 Estimate for the Year 2000
Total World Population	7,400,000,000 or 100% increase over 1972
Total U.S. Population	264,000,000 or 26% increase over 1972
Under 18 years of age	Increase of 3,894,000 or 5%
20–34	Increase of 9,353,000 or 21%
35–49	Increase of 26,063,000 or 75%
50–64	Increase of 8,389,000 or 27%
Over 65 years of age	Increase of 7,893,000 or 38%
Rural Population	Decline due to migration and decreased fertility
Urban Population	Increase due to migration to urban centers
Public School Enrollments	Only slight increase due to decreased fertility
College Level Enrollments	Significant increase due to older population and changing technology
Proportion Females Employed	Increase due to decreased fertility
Health Care Needs	Increase due to improved standards
Retirement Age	Older due to higher pension costs
Population Trend	Negative growth due to continued decline in fertility (Estimated leveling in U.S. at 300,000,000 sometime in the twenty-first century.)
Income Trend	No relative differences currently indicated. Eight to 10% at submarginal level. Approximately 1/3 of population at marginal level, and 50% at middle or upper level.
Family Size	Decrease at lower income levels and increase at upper levels

systematic methodological approaches are available for arriving at reliable group consensus.[6] Human-service system requirements, as projected in the present chapter, however, are based on a brief review of the literature, and are not taken to be conclusive nor necessarily representative development trends.

SUBSYSTEM DEVELOPMENT TRENDS

Development trends in the total human-service systems complex may continue largely in the process of random evolution for the foreseeable future, with both the public and various political factions posing formidable elements of resistance to systematic and planned change.[7] Natural forces of population pressures, economic survival needs, fuel and energy shortages, depletion of natural resources, contamination of air and water, etc., may necessarily militate for changes in the human-services delivery system and its subsystems. Table 7 projects a number of human-service subsystem requirements over the next several decades based on rational considerations.

Domiciliary Subsystem

With increasing emphasis on the importance of the environment on health, behavior, and quality of life, general improvements will continue to be sought in living-space design and structural utilization.[8] Large-scale studies of housing projects (prefabricated and individually constructed) and neighborhoods should be increasingly undertaken, calling for human-factors and systems expertise, particularly in considering health, welfare, safety, convenience and housing quality, transportation and facilities quality and efficiency, and other more totally interrelated elements of units and neighborhoods. Requirements studies must involve neighborhood aggregates as well as individual family units.[9] Requirement parameters for total neighborhoods might include:

—constituency by family size, race, age, and income
—political affiliation
—social mobility
—levels of activity
—employment levels
—stages in life cycle
—social group involvement
—vulnerability to natural disasters, both acute (e.g., earthquakes, floods, earth slides, brush fires, etc.) and chronic (e.g., radiation fallout, smog, noise, etc.)

Requirements for individual family units may relate to:

—family size, income and constituency[10]
—life style as defined by income, stage in life cycle and marital situation
—unit dwelling and family interactions (time spent in different activities by location)
—bathroom (elimination, laving and personal hygiene, reading, etc.)
—bedroom (sleeping, cohabitating, reading, watching television, etc.)

TABLE 7
Projected Human-Service Subsystem Requirements

Human Service System	Population Served	Projected Requirements
Domiciliary	Low middle- and lower-income groups, particularly high-density low-income groups, or approximately 10% of total population comprising the latter.	Generally improved space and environment utilization. Resourceful low-cost housing for low-income groups. Criteria include access to work, school, and commerce, fire and crime protection, sanitation, air and noise pollution control.
Industry & Commerce	As source of employment and consumer products for low- and middle-income groups (ca., 80% of total population).	Improved product design, and planning consistent with population growth, ecological and energy constraints. Criteria include improved access, optimal man-machine processes, noise, odor, air and water pollution control.
Surface Traffic Control	Primarily urban and interurban estimated to become between 70–80% of total population at turn of century.	Improved and innovative mass transit systems; curtailment of private auto usage. Criteria include decongestion, safety, increased volume of passenger flow, passenger acceptance and comfort, rapid ingress and egress, convergence and divergence.
Air Traffic Control	Increased air travel is expected by lower-income groups, e.g., while it is now estimated that 15% of upper-income groups use 80% of air service, this should become 40–50% of lower-income levels contributing to an estimated 30–40% increased volume of air travel.	Improved air terminal access from inner city; improved processing for boarding and air piracy control; air and noise pollution control; improvements in safety in takeoff and landing and enroute control operations, and efficiency with relieved workload on traffic controllers. Continuing improvements in passenger comfort and flight safety.

TABLE 7 (CONT'D.)

Human Service System	Population Served	Projected Requirements
Fire Protection	Primarily high density population in inner city.	Improved efficiency in allocation of man-machine functions. Operational determination of highest risk and cost areas in deployment and damage. Improved rescue functions; increased control of false alarms and arson.
Law Enforcement & Criminal Justice	Crime rates highest among youth; some decline in this population indicated, though increased density. Civil cases likely to increase appreciably with rise in older adult population.	More emphatic efforts in improving preventive measures. Improved man-machine functions allocation in adjudication, eliminating processing delays; improved and/or reformed corrective rehabilitation, with efficient linkages developed to health, education and welfare.
Sanitation	Inner city, low-income, high-density population (approximately 30% of total population).	Improved processing of solid, liquid and gaseous wastes through neutralizing, recycling, safe disposal, and accelerated decomposition. Improved allocation of functions and increased handling efficiency. Control of radiation hazards with increasing use of nuclear energy sources.
Health Care	Low, and marginal income, and elderly population comprising approximately 35% of total population.	Improved general health care through increased efficiency in utilization of resources and processing of outpatients. Use of out-patient services substituting for in-patient; studied allocation of functions for machine processing and various professional and subprofessional levels. Improved equipment and machine interfaces in diagnosis and treatment.

TABLE 7 (cont'd.)

Human Service System	Population Served	Projected Requirements
Mental Health	Low socioeconomic groups, or approximately 20% of the population, will use approximately 80% of the public mental health resources.	Improved multicoupling of services with decline of medical model. Design of open-loop health, education and welfare services for prevention and rehabilitation.
Public Welfare	Categorical assistance provided to approximately 10% of the total population (5% at any one time).	Change in basic policy formulation from sustaining support to productive rehabilitation. Design of open-loop health, education and welfare services. Improved processing through appropriate, efficient operations and functions allocation.

   —maximal and minimal occupancy relative to mental health and physical illness

   —a taxonomy of activities as criteria for layout, facilities and convenience

Current housing in the unplanned, ill-conceived and poorly designed mode has been described as lacking specified functions, having over-designed or inappropriately allocated spaces, lacking equilibrium in various waste and environmental control aspects, inefficiently utilizing energy, poorly processing information, economically burdening the constituency with unattractive structures and landscapes, etc. Future approaches, through human-factors and systems engineering, may serve to mitigate such gross areas of deficiency in the domiciliary subsystem.

Industrial and Commerce Subsystem

The private enterprise aspects of this subsystem harbinger well for continued improvements bearing on production efficiency and cost saving. However, the human-service elements, unless specifically relevant to survival and prosperity of the individual industrial or commercial component, may frequently be neglected or even costwise at odds with more optimal solutions, often requiring government intervention. Among requirements for the future will include:

—improved product design for human operation, safety, maintainability, etc.

—enhanced efficiency in energy consumption and the use of alternative fuel sources

—improvements in the use of human operators in production operations, to avoid dehumanizing effects, etc.

—location and development of industrial and commercial facilities for improved access by workers, shoppers, etc.

—vastly improved air, water and noise pollution control and recycling of solid waste

—improved vehicular and pedestrian control in the layout of shopping plazas, store interiors, etc.

Surface Traffic Control

At this writing, the high-rise Sears- & Roebuck Building on South Wacker Drive in Chicago is being occupied. Executives, managers, office workers, etc., numbering from 16,000 to 18,000, will fill this tallest structure in the world over a land area of less than one city block. Each day, about the cluster of high-rise office buildings within two or three blocks in the South Wacker area of Chicago, 40,000 to 50,000 persons will converge from commuter trains, buses, elevated railroads, and to a limited extent in this inner city area, automobiles. Each night they will seek the same conveyances to diverge from the area to outlying suburban areas or high-rise apartments within the city. It remains to be seen to what extent congestion, delays and inefficiencies will plague this new operational complex; but the already overloaded elements of the transportation system can promise only to contribute to the growing problem of urban transportation. Mass transit components of the surface traffic control subsystem may become increasingly involved in solution of inner city problems, but will require extensive operations research and human-factors studies relative to the total system needs. Requirements and constraints in such design studies may be identified as follows:

—efficient interfacing with the total transport system (land, air, sea, pedestrian, etc.)

—system flexibility and adaptability (expansion, reduction, disbandonment, etc., depending upon total configurational requirements)

—safety practices consistent with performance requirements and operating methods (e.g., not to exceed 0.15 fatalities per 100 million passenger miles)

—designed in accordance with specified conditions and tolerance for operation in normal, extreme and emergency weather and environmental conditions

—fail-safe design features, i.e., failure shall revert the system to a state known to be safe. Indicators and safety interlocks shall intrude on the operator's functions to cause a safe condition in failure

—human error propensity to be minimized in operation, assembly, and maintenance

—reliability standards for equipment and human operator

—maintainability—with built-in test provisions, minimum specialized maintenance skills, stress derating, redundancy, modular packaging, access, cannibalizing provisions, etc.

—passenger comfort and convenience, e.g., in seating, vibration, acceleration, time saving, etc.

—efficiency in queuing, boarding, detraining, etc.

—control of public using behaviors, i.e., trade-off with convenience, privacy, costs, etc.[11]

Air Traffic Control

Among the major system and human-factors requirements in future air traffic usage may be simply control of increased passenger volume over long-distance travel. This may be somewhat alleviated through competitive intermediate-range surface transport systems, particularly when total trade-off in time and cost comes generally to favor the surface modes. Future analytic and operational requirements will likely include the following:

—extensive operations research studies in designing alternative modes of travel for various populations, e.g., workers in commerce, industry and government versus tourists, etc; travel radii by subgroup, etc.

—improved access to air terminal from inner city and other high-density points of departure

—improvements in passenger routing at the terminal to flight line, the screening process and boarding control procedures

—improved functions allocation and tasks design in data and communications on incoming flights, flight plan clearance, outbound and inbound ground control, and take-off and landing control

—improved provisions to minimize environmental pollution by noxious exhaust chemicals and noise

Fire Protection

As in almost all areas of human service, major requirements are indicated for analytic and operations research studies in fire protection. We indeed need to know high-density and high fire risk areas, elements of highest cost and inefficiencies, the most effective deployment of men and machines in total mission sequences, etc. The following may be identified as several areas for design improvements:

—revision of fire insurance laws to negate incentives for arson
—expedited slum clearance and elimination of substandard housing to avert other incentives for arson and harassment of the fire-fighting forces
—improved cost benefits through allocation of man-machine functions (over 90 percent of costs are currently attributable to salaries)
—improved technology and man-machine interfaces in tactical communication, ventilating tools, fire detectors, extinguishing agents, alarm transmissions, etc.
—improved site location of resources based on incidents and reaction-time requirements

Law Enforcement and Criminal Justice

Deficiencies in both law enforcement and criminal justice have been acknowledged for several decades. The general public has been concerned lest they themselves or their property may be violated or confiscated. The recidivism rates have also been generally alarming, while inefficient adjudication functions have been crystallized into the folk shibboleth, "Justice delayed is justice denied."

Several major requirements over the next several decades may be stated as follows:

—improved methods of prevention or deterrence through operations research on high-risk groups of offenders
—improved methods of detection and apprehension
—appropriate functions allocation in adjudication minimizing reaction time, e.g., machine processing of standard violations with corrections-specific measures of restitution
—improved functional legislative definitions of criminal and civil offenses
—corrective, retraining or changing orientation indicated through environmental design and improved practices of incarceration institutions

Sanitation

Requirements in the sanitation industry will likely relate to continued improvements in environmental protection and improved processing and control of health hazards. Generally, these may include the following:

—provisions for improved control of litter about domiciles, on highways, etc.

—improved garbage and refuse methods of handling

—operations research studies to establish areas and density of waste accumulation by material, i.e., solid, liquid, or gas

—identifying requirements and improving technology in recycling, and accelerating functional decomposition

—improved technology in projecting health hazards and epidemiological implications of various environmental practices

Health Care

Health care requirements for the next several decades will undoubtedly focus increasingly on the older and middle-age groups. (See Table 7.) This area, too, seems to have been neglected in meaningful operations research (OR) studies, specifically establishing unfilled health-care needs by age, income, etc., subgroups. Given such OR data, appropriate functions analysis might be completed for more optimal and efficient allocation of health resources. Future requirements might concern the following:

—integration with more comprehensive health and welfare needs

—more functional in-patient or hospital usage in conjunction with improved out-patient services

—improved manpower functions relating to required tasks of care, i.e., in utilizing automated design and use of paraprofessionals or subprofessionals

—improved funding mechanisms circumventing vested professional medical interests

—improved preventive health care through education and research

—improved diagnostic technology through improved instruments and efficient design of man-machine interfaces

—improved health-care environmental design for patient care and access by professional and subprofessional staff in emergency and routine treatment

Mental Health

This human-service subsystem has been largely dominated by a medical approach to an ill-defined population of needs. The needs, however, have more often related to training rather than medicine, for correcting behaviors and restoring or habilitating the patients for a useful, functional life. Several requirements may generally be projected as follows:

—integration with more comprehensive health and welfare needs
—improvements in evaluation methods
—identifying functional medical applications, with trade-off based on alternative approaches
—increased efficiency in the use of computer functions based on information requirements analysis
—effective systems design for rehabilitation of chronic mental patients
—increasing use of day-care alternatives to in-patient approaches to treatment
—improved incentive structures for mental-health care givers oriented to patient improvement

Public Welfare

Some precedence is indicated in the study of this human service subsystem exploiting human factors technology.[12] However, continued improvements in methods of study are needed in identifying the more inclusive populations at risk, while relating these to system development goals and alternative approaches to service.[13] Several human-service requirements may be projected for this subsystem as follows:

—integration with more comprehensive health and welfare needs
—shift in emphasis from simple sustenance to rehabilitation, habilitation, or development of productive life styles
—development of simulation models for meaningful interfacing of public administrative personnel and elected officials for testing such alternative service modes as guaranteed income, various rehabilitation models, etc.
—increased emphasis on incentive structures for service recipients
—increased emphasis on incentive structures for service staff

PROJECTED CONSTRAINTS AND
BUREAUCRATIC RESISTANCE

Projecting human-service requirements for the decades ahead is at best a tenuous undertaking. It is, of course, quite necessary to posit such an outlook in order to gain perspective on the prospective employment of systems and human-factors technology. However, the inertia inherent in bureaucratic and practical day-to-day operations, as well as in short-sighted political expediencies, seems generally to militate against planning for projected outcomes. New housing programs are stymied when we consider what to do with present occupants of slum housing. Commercial and industrial progressive long-range planning is overridden by short-term profit-taking considerations. Indeed, any planning that must disrupt the current operational modes, displacing those of us in immediate employment, interrupting our income, etc. (which in itself may be attributed to failure in planning) would seem most naturally to be resisted.

Rather, the system, in lieu of planning, seems to be one of reacting to crises. We suffer delays, congestion, and exhaust fumes when commuting to work, for this is "the price of progress and a highly productive industrial society," etc. Pressures are brought to bear by current service and equipment producers when the road to the airport, and the airport itself, is overrun with people. Their answer is in additional limousines and taxis. Fire-fighting improvements are resisted by the fragmented and political-lobbying supply industry. The police and justices may be most reluctant to yield an essentially controlling power base. In the face of an imminent fuel crisis, political expediency in the popular mode dictates environmental compromise, rather than accepting the environmental constraints as given. Again, political and vested-interest expediencies, rather than real system solutions in health, mental health and welfare, tend to dictate operations in these human-service areas.

Indeed, planning or projecting the human-service requirements of the future is so bound up in the network of the political and vested-interest complex that these must be inherently considered part of the planning project. The key to such future planning may thus very well lie in the effectiveness with which functional alternatives and practical consequences can be communicated, not only to public administrators but to the public.

NOTES TO CHAPTER 7

1. E. Weissmann, "World Cities in the Future—Shelter Systems for the Future," (Paper presented at the 137th Annual Meeting of the American Association for the Advancement of Science, Chicago, Ill., December 28, 1970); K. Watt, "Will

the Future Be Shaped by Rational Policies?", *Saturday Review*, October 24, 1972, p. 76 f.

2. E. Banfield, *The Unheavenly City* (Boston: Little, Brown and Company, 1970).

3. T. Rowan, "The Role of System Science in Modern Socioeconomic Problems," (Testimony given before a subcommittee of the Senate Labor and Public Welfare Committee, chaired by Senator Gaylord Nelson of Wisconsin, November 19, 1965); I. Hoos, "Systems Analysis as a Technique for Solving Social Problems—A Realistic Overview," Working Paper No. 88, Space Sciences Laboratory, Social Sciences Project (Berkeley, Calif.: University of California, 1968); R. Lynton, "Linking an Innovative Subsystem into the System," *Administrative Science Quarterly* 14 (September 1969):398–416; J. Bower, "Systems Analysis for Social Decisions," *Operations Research* 17 (November 1969):927–40.

4. R. Chartrand, *Systems Technology Applied to Social and Community Problems* (Rochelle Park, N.J.: Spartan Books, 1971).

5. "Population Slowdown—What It Means to U.S.," *U.S. News and World Report*, December 25, 1972, pp. 59–62.

6. N. Dalkey, *The Delphi Method: An Experimental Study of Group Opinion*, Memorandum RM 5888-PR (Santa Monica, Calif.: The Rand Corporation, 1969); S. Umpley, *The Delphi Exploration. A Computer-Based System for Obtaining Subjective Judgments on Alternative Futures*, Report F-1 (Urbana, Ill.: Computer-Based Education Research Laboratory, University of Illinois, 1969).

7. E. Lindemann, "Social System Factors as Determinants of Resistance to Change," *American Journal of Orthopsychiatry* 35 (1965):544–57.

8. J. Wohlwill, "The Emerging Discipline of Environmental Psychology," *American Psychologist* 25 (April 1970):303–12.

9. A. Wilson, "Modeling and Systems Analysis in Urban Planning," *Nature* 220 (December 7, 1968):963–66; D. Carson, "Human Factors and Elements of Urban Housing," in *Human Factors Applications in Urban Development* (New York: Riverside Research Institute, 80 West End Avenue, 1970).

10. U. S. Bureau of the Census. Census of Population: 1970 detailed Characteristics. Final Report PC (1)—D1 Social and Economic Statistics Administration. U.S. Government Printing Office, Washington, D.C., 1973.

11. W. O'Connell, *Ride Free, Drive Free: The Transit Trust Fund and the Robin Hood Principle* (New York: John Day Company, Inc., 1973).

12. Space-General Corporation, *Final Report: System Management Analysis of the California Welfare System* (El Monte, Calif.: 9200 East Flair Drive, 1967).

13. "Recounting the Poor," *Trans-Action*, November–December 1966, pp. 6f.

8

Functions Analysis and the Open-Ended Means Approach to Human Services

If progress is to be maintained in the design of human-service systems, executives, legislators, administrators, and, indeed, the public itself must be continually convinced—not only of the need for ongoing improvements, but of the most economical and effective means to accomplish such improvements. This might mean, for creative and innovative purposes, the assumption of a completely open stance in considering variable means to accomplish human-service subsystem functions.

L. L. Thurstone in his published work, *The Nature of Intelligence*, during the twenties saw the capability to suspend the selection of means as characterizing the very nature of human intelligence. Lower animal forms, he theorized, closed rapidly on a means solution—to run after and pounce on their prey, only to remain chagrined and panged with hunger when the prey found easy avenue of escape. The intelligent animals, according to Thurstone, were those who paused to consider, reflect, and project outcomes for various alternative possibilities. This, indeed, he postulated, was what separated man from the animals or the intelligent from the less intelligent.

Modern technical experts have indeed applied such theoretical intelligence postulates to an open-ended means approach in design or action solutions.[1] Such analyses, however, have been largely conducted in military or industrial contexts, while the entire arena of human services appears to have been largely means-bound without benefit of a search for alternatives.[2]

In the present chapter, several areas of technology will be considered which could promote the search for alternative solutions in human services; these particularly concern the methods of functions analysis and operations research as they bear on the testing of various means solutions before actually committing major resources to often irreversible processes.[3]

136

FUNCTIONS ANALYSIS

System or subsystem functions, as described earlier, are derived from mission analyses following policy and SOR specifications. Identified functions may be broad and general, such as maintaining a comfortable environment; or they may be more specific and detailed, as in providing temperature regulation at 70° F. within ± 2°. However, at the functions level of analysis specific means are kept open until trade-off criteria are developed to facilitate decision making. The cost and sensitivity of sensory elements, such as mercury or alcohol, for example, may then be introduced for consideration before arriving at a definitive decision on the means to achieve temperature regulation. Thus, the means are open-ended, while extensive and often complex operational studies are needed to arrive at better or best design solutions.

In human-service subsystems, we are most frequently means-bound, having failed to consider functions at the outset that might be accomplished by any of a number of means, some of which are less costly and more effective than others. During the analytical process it becomes necessary to make such determinations for each function within the subsystem. The analytical technique employed in making such determinations is called "Operations Research," essentially comprising an open-ended means approach.

OPERATIONS RESEARCH

The modern technology of Operations Research originated in Britain during the early years of World War II when limited military resources required optimal efficiency in deployment for national survival. Kill probabilities of depth charges in antisubmarine warfare, optimal dispersal of British air power in the air war, etc., were analyzed to derive the greatest effectiveness against the superior numbers and military might of the German war machine.

After the war, several soundly managed commercial organizations applied Operations Research to business and sales operations. The method has since been applied extensively in a number of industrial management and marketing areas, while an effort has also recently been made to employ the technique in several human-service areas.[4]

The essential element in Operations Research problems is the existence of alternative courses of action with a choice to be made among them. Goals or design objectives and outcome criteria measures are also inherently required in the analysis. The method is chiefly employed to clarify the relation between several courses of action, determine outcomes in cost and service efficiency, and establish which course best relates to the accomplishment of overall policy goals of the system.

As a tool, an Operations Research model may be described as exact or probablistic. The exact model is one where effects may be determined within fairly precise limits, as in standard production runs, serving lines in a company cafeteria, etc. Probabilistic models, on the other hand. contain major elements of uncertainty, as when indeterminate consumer behaviors are involved, or changing requirements for welfare services, variations in the crime rates, or the economy, etc. The use of probabilistic models does, however, permit a faithful representation of total actual operation, for example, the linking of sales, promotion, production, and distribution operations. The analysis requires manipulation of the model to determine outcomes and other parameters upon which to predicate decisions. While the manager is therewith provided a reasonably accurate tool to determine likely consequences of various decisions, the responsibility for these decisions is his own and can in no way be supplanted by the analysis alone.[5]

Operations Research is, of course, part of the engineering or design process, but the Operations Researcher differs in his approach or focus from that of the engineer. A design engineer, for example, may be oriented to the development of increased explosive power; the Operations Researcher in its effective deployment relative to depth or height for detonation effectiveness in kill probability. The engineer will design a communication control system to get clear and sufficient information quickly to the railway operator; the Operations Researcher, in contrast, approaches the problem from the standpoint of whether increased speed and clarity of information will contribute to the trains arriving at their destination safely and on time. The engineer may troubleshoot equipment breakdown and seek means to prevent it; the Operations Researcher analyzes the problem in terms of optimizing the operation with present equipment and the relationship between equipment breakdown and its critical function in the system.

Thus, the employment of Operations Research techniques may best relate to functions as they are delineated within the subsystem. Operations Research analyses may contribute most meaningfully to the design decision in clarifying alternative solutions from a cost and effectiveness standpoint, to accomplish system and subsystem objectives.

OPERATIONS RESEARCH IN HUMAN SERVICES

Exercises in the assessment of alternatives for human-service system design may increasingly call attention to the inadequacies of current operations. Table 8 presents illustrative Operations Research problem definitions for typical subsystem functions. Operations Research (OR) problems are necessarily bounded by realistic contraints, which must be considered in the OR solution or, indeed, as implicit goal criteria by which given alternatives might be evaluated. In each case, common constraints and system linkages are specified for each subsystem function, with a typical type of OR problem structured for each function, i.e., analysis of alternative design solutions.

Domiciliary Subsystem

Functions in the domiciliary subsystem include internal environment control (light, temperature, humidity, sanitation, insect and vermin control, etc.,), water, sewage disposal, and living density limitations in privacy and personal space. A typical OR problem for such functions may include assessment of alternative transparent area design or exposure to various solar angles, etc., using fuel economy, personal reaction, etc., as design criteria. Similar OR problems may be defined for external environmental hazards, the juxtaposition of living areas, etc.[6]

Commerce and Industry

Numerous OR problem definitions may be developed within the context of commerce and industry for such functions as site location, product planning, fuel and energy conservation, product diversification, pollution control, marketing, shipping, etc.

The Surface Traffic Subsystem

Surface traffic functional alternatives for OR assessment may range from trade-off studies of current modes compared with those of completely new design, e.g., magnetic levitation, to the value of improving and integrating all available current modes of travel.[7] Major current functions, of course, relate to minimizing energy and fuel consumption.

TABLE 8

ILLUSTRATIVE HUMAN-SERVICE FUNCTIONS AND OPERATIONS RESEARCH PROBLEM DEFINITIONS

Subsystem	Function	Constraints and Linkages	Operations Research Problem Definition
Domiciliary	Internal environment control	Fuel and energy limitations, and costs and acceptance.	Variable window exposure area and insulation characteristics versus cost, fuel and energy consumption, and personnel reaction/performance.
	External environmental hazards control	Relocation costs, and resistance of home owners, access to work in new location, etc.	Trade-off studies in damage, etc., for periodicity of flooding, earth slides, fires; restoration in place as against relocated construction.
	Complexity of envelope	Space limitations, costs, functional usage, standardization and personal reaction.	Vertical, diagonal, horizontal, etc., variable juxtapositioning of living and functional areas versus cost, access, personnel reaction/performance, etc.
Commerce and Industry	Fuel and energy conservation	Projected availability, costs and functional priorities in goods and services, unemployment, etc.	Operational and timely productive capability of various energy source modes, development time and costs, versus safety, production and employment level, etc.
	Pollution control	Projected epidemic health problems within time limits, costs, employment, aesthetic character, etc.	Projected time to control development within criteria limits for various chemical filtering etc., processes against costs, side effects residue, effect on employment level and production schedules, etc.
Surface Traffic	General mass movement in and out of inner city.	Extant private motor vehicle predominance; fuel limita-	Comparative studies of mass transit modes versus current modes against congestion, speed, operating

TABLE 8 (CONT'D.)

Subsystem	Function	Constraints and Linkages	Operations Research Problem Definition
Surface Traffic (cont'd)		tions; areas of population concentration; costs; employment levels and retooling for change; passenger acceptance.	and development costs, etc. Trade-off studies among various mass transit modes relative to energy requirements, costs, efficiency, noise, passenger acceptance, etc.
	Transfer or translocation to specific inner city sites.	Concentration of inbound and outbound population by time schedule; vertical and horizontal layout of buildings; costs; congestion; aesthetic appeal; convenience and acceptance.	Trade-off studies on private vehicles versus walk conveyor installations, shuttles, etc., against criteria of operating and development costs, efficiency, passenger acceptance, etc.
Air Traffic	Mass traffic convergence to and divergence from air terminal.	Extant roadways, private autos, taxis, limousines, etc.; congestion; costs; fuel conservation; efficiency; passenger acceptance.	Trade-off studies of various conveyance modes and dispersal functions—substations with mass transit runs to air terminal and prohibition of road traffic, versus current modes against operating and development costs, speed, efficiency, passenger acceptance, etc.
	Air traffic terminal control.	Conventional, vertical and short take-off aircraft, extant skills and employment in traffic control, etc.	Automated control procedures versus present control modes against development and operating costs, efficiency, safety, skill development and employment opportunities.

TABLE 8 (CONT'D.)

Subsystem	Function	Constraints and Linkages	Operations Research Problem Definition
Fire Protection	Prescreening of fire severity and priority.	Rapid response required; filtering false alarms; priority in control by property value; information system minimal.	Various alternatives to be assessed—radio dispatched coordinates to prowl car for rapid response in screening and immediate preliminary action. Trade-offs with current and other modes against reaction time, safety, operating costs, damage, etc.
	Fire extinguishing	Speed, safety, minimal damage, spread control, costs, etc.	Automated functions built at site versus manual reaction from station against operating costs, effectiveness in magnitude of damage, personal safety, etc.
Law Enforcement and Criminal Justice	Identifying offense or criminal violation	Conventional definitions of crime; public safety margins; costs; quality of life; political pressures, etc.	Study of effects in extending dimensions of tolerance for misdemeanors, felonies, no-fault offenses, etc., versus current arrest practices against cost, manpower, deployment, recurrence of act, etc.
	Apprehension	Current police methods, practices and orientation in physical approaches, public safety; police safety; costs; recidivism; political and social pressure, etc.	Apprehend on the basis of guilt manipulation making the culpable aware of his offense and required restitution compared with current system on the basis of cost, public safety, recurrence of offense, etc.
	Adjudication	Political pressures; public safety; vested interest of justices, etc.	Standardized computer programs for automatic adjudication versus current processing against speed, cost, public safety, public acceptance, etc.

TABLE 8 (CONT'D.)

Subsystem	Function	Constraints and Linkages	Operations Research Problem Definition
Sanitation	Garbage collection	Maximally unobtrusive; costs; noise; odor; air pollution; vermin control; population density, etc.	Automatic versus manual pickup processes for high density population against cost, efficiency, reliability, safety, vermin control, etc.
	Disposal	Costs; noise, odor; pollution; recycling efficiency; fuel and fertilizer products; aesthetic degradation, etc.	Dispersed or localized incinerator or recycling operation versus centralized processing against costs, efficiency, noise, aesthetic, pollution, etc., control criteria variables.
Health Services	Diagnosis	Medical vested interests and control; patient safety; reliability and validity; costs manpower availability; patient density, etc.	Automated sensing and computerized programming or semi-automated functions versus current manual medical processing against costs, reliability, patient safety and acceptability.
	Medical ministration	Medical prestige and regulation of standards; patient safety and effectiveness; costs; manpower availability; reliability and validity; etc.	Standardized semi-automated prescription program from formated data base versus standard medical practices against criteria of effectiveness, costs, manpower availability and training requirements, etc.
Mental Health	Processing first petition for certification of mental illness.	Conventional medical definitions; judicial convention regarding adjudication of mental illness; medical and related mental health vested interests; public regard and	Processing through a day-care service rather than for full-time hospitalization; comparing costs, rates of return to full-time employment, relapse rates, etc. Devise an education and training model for com-

TABLE 8 (cont'd.)

Subsystem	Function	Constraints and Linkages	Operations Research Problem Definition
Mental Health (cont'd)		convention; integration of multi-agency services; costs; effectiveness, etc.	parison with the current psychotropic medication, psychosurgery, electroshock, etc, medical model. Compare on the basis of costs, patient social and vocational skills, relapse rates, etc.
	Sustaining care of patients discharged from mental hospital.	Prolonged stigma; public attitude; limited funding; programs for service ill-defined; vested interests of hospital factions, etc.	Identify discharged population at highest risk of relapse or return to mental hospital and deliver appropriate or effective services while reacting to feedback on costs, social occupational adjustment, readmission rates, etc.
	Patient billing	Low socioeconomic population served; harassment of nearest relative for payment; cost of billing process; net income derived, etc.	Compare value derived from billing lower socio-economic segments versus no billing.
	Staff activity reporting	Accounting for services provided demanded by legislature and community boards; staff time consumed in reporting, etc.	Compare computer versus manual processing; compare PERT and other techniques of program monitoring, etc.
Public Welfare	Administering sustaining-care to the poor.	Dysfunctional public attitude toward "being poor;" vested interests of resource-holding	Compare a vocational development model with current minimal support practice in welfare rights

TABLE 8 (CONT'D.)

Subsystem	Function	Constraints and Linkages	Operations Research Problem Definition
Public Welfare (cont'd)		groups; stigma of public socialization or state support; program costs; inherent political conservatism; racial prejudices; costs; acceptable effectiveness measures, etc.	against costs, outcome, sustained independence, child development criteria, frequency of fraud, etc. Compare guaranteed annual income method, varying minimum level, with current methods of support and above model, against costs, health criteria, independence, child development, etc.

Air Traffic

Typical current functional problem areas for OR study include speedy access to commercial take-off and landing sites. Innovative and creative OR studies are needed in this area, such as dispersed subterminals connected by rapid transit systems versus remote centralized processing, etc. Trade-off criteria may be readily obtained in such measures as time, efficiency, public acceptance, etc.

Fire Protection

Multiple functional areas in the fire protection subsystem lend themselves nicely to OR analyses.[8] The central OR problem appears to be one of limited resources to be most effectively applied against such effectiveness criteria as death and injury rates, property damage, operating costs, optimal monitoring and prescreening, etc.

Law Enforcement and Criminal Justice

Design criteria for OR trade-off studies in this subsystem appear to be fairly clear cut—protection of society, minimizing crimes of violence and property, socializing of offenders and reducing their rates of offense, etc. However, constraints imposed by the momentum of current operational modes, and the traditional reaction to crime must be considered as part of the OR problem—if not as measurable objectives then as obstacles to be overcome in public education programs or through news-releases on results of studies and recommendations. Here, too, creative OR studies are needed to assess radical new approaches to this human-service subsystem.[9]

Sanitation

Functions amenable to OR type analyses include population density and waste volume studies for assessment of new alternative approaches to garbage collection, recycling waste products for fuel production or reusable solids, and liquid and air pollution control.

Health Services

Major interest is evident throughout the country in developing innovative programs in the health care subsystem.[10] In fact, health appears to be one of the most advanced areas in sophistication advocating the use of Operations Research techniques,[11] though concrete applications are still wanting. Health care functions that could meaningfully lend themselves to OR-type analyses include prevention, diagnosis, medication ministration, prescription, surgical repair, convalescence and hospitalization.

Mental Health

The mental health subsystem is only slowly turning to such methods as Operations Research for analysis, perhaps due to the centuries-old force of conventions regarding mental illness, and the dominance of the medical model. However, new, though fragmented, interest is awakening in the employment of such analytic methods.[12] Functional areas amenable to OR-type studies include alternatives to hospitalization, high-risk population segments where maximal results may be achieved for minimal effort, alternative methods of dispensing medication, alternatives to blanket patient billing for services, and alternatives in accounting for staff time.

Public Welfare. New programs are often proposed or introduced in the public welfare subsystem.[13] However, available analytic methods in this human-service area have been only minimally employed, perhaps partly because of lack of hard data for evaluative trade-off studies. Possible OR studies include variable levels of guaranteed annual income, alternatives to eligibility processing, introduction of rehabilitation-oriented objectives, etc.

A HUMAN SERVICES OPERATIONS RESEARCH STUDY—AN EXAMPLE

OR studies may often attempt to encompass the dynamic relationships among global, complex and inclusive linkages in the system;[14] or they may relate simply to several isolated key parameters. The latter type model thereby may provide essential information on such variables as costs and numbers served, while value, benefits or effectiveness of service must be established as an adjunctive study. The chief advantage of such simplex OR models is that they may be more easily understood by administrators, be less intimidating, and more readily explained to legislators and the public than are the more technical and complex models. Thus, the raison d'être of the OR analysis is retained, viz., to consider alternative design solutions, while bringing it within the scope of other than the OR technical experts to consider.

Such a simplex OR model was developed for the alternative of partial hospitalization of first-time-admission mental patients. Evidence was accumulating that over 90 percent of such patients, with the full range of major psychiatric diagnoses, could be more effectively treated in day hospitals, returning home or to another place of residence at night. They thereby did not assume the sick role associated with hospital beds, returned to their normal job situations more rapidly, required less overall treatment time and only half the relapse rate of the randomly hospitalized inpatient.[15]

Based on actual costs for such inpatients hospitalized from a 16-county area in east-central Illinois, a trade-off alternative study was made to determine costs if day care had been employed for these patients. The costing formula developed was as follows:

$$x = (a \times b \times c) + y (a \times b \times c),$$

where x was total annual costs; a, the number of patients; b, their average number of days in treatment; c, the average cost per day; and y, the proportion of the total group readmitted for treatment within the year.

Eliminating those patients with legal involvement about whom questions of "danger" might be raised, the following per annum volumes were indicated:

> To state hospitals—37
> To regional centers—67
> To general hospitals—2964

Day hospital costs were derived from day hospital charges made by one local general hospital, while other costs were actual. Cost savings indicated by substituting day hospitals for this population were $198,191 for the state hospital patients, $200,152 for those in regional centers, and $1,916,510 for patients sent to general hospitals, for a total cost saving of $2,314,853 per year. This cost saving was quite dramatic, particularly when the alternative studied presented most favorable evidence as being superior to the traditional modes of service.

The data from such a simplex OR model was most easily dispatched to administrators and a news medium, with several levels of administration responding enthusiastically; whereas a more complex OR model may have been less communicable to the administrators, this simplex model began to make sense to them.

FUNCTIONAL ALTERNATIVES INVOLVING MANPOWER

Throughout the United States today it is estimated that over half the total manpower complex is now producing services rather than goods or hardware products. In the human-service industry this must be indicated at well over 90 percent, with only minimal production of goods in such subsystems as sanitation which produces fertilizer and fuel from liquid organic wastes, etc.

Manpower specialization in such a vast manpower pool cannot be taken as a major limiting constraint in the selection of functional human-service alternatives. Such constraints may well be a kind of myth perpetuated by vested professional interest groups in medicine, social work, law, etc. Rather, from a systems standpoint, virtually any normal human operator may serve many of the same functions as paraprofessional workers with only nominal qualitative and quantitative skill requirements. Specific system operational alternatives may simply require a detailed taxonomy of task elements to be derived, from which skills and training requirements may be determined as the substantive functional constraints in the system.[16]

EFFECTIVENESS EVALUATION

Each proposed alternative means by which to serve human-service subsystem functions must necessarily be infused with, or contingent upon, explicit measures of outcome or goal accomplishment. This, of course, might presuppose the operation of an ongoing data base, or at least the availability of correlated outcome data associated with each alternative. These data may often not be readily accessible. For example, assessing cost damage from floods or earth slides may require historical research. Traffic congestion measures may require special instrumentation or systematic observations. Passenger, patient, or general public satisfaction and acceptability of a service mode may require much depth and breadth of probing to acquire the essential data base for comparison. In fact, virtually all OR analyses in human-service subsystems may require an essential research component specifically directed to assessing comparative design solutions.

Effectiveness evaluation must necessarily be oriented to policy-directed outcome measures. Design for this method was discussed earlier, as illustrated in Figure 6. Indeed, without the feedback loop essential to goal accomplishment, administrative control is attenuated often to the point of sterility. President Kennedy, during the Cuban missile crisis of the 1960s, was advised by the Russian Premier Khrushchev that the United States must dismantle its missile bases in Turkey as a condition for Russian withdrawal from Cuba. Kennedy had much earlier ordered such dismantlement, but, lacking an essential feedback loop or progress-reporting system, it remained for the potential enemy and foreign power in confrontation to advise him of his administrative failure.

Aggressive, goal-directed administrative authority is, of course, essential to the ultimate fruition of OR analytic power. In the human-services area, such administrative authority is partially or wholly contingent upon political and public support.

Through functions analysis and the open-ended means approach, predicated upon OR analyses and intuitive scholarship, an enlightened administration of human services becomes possible; thence, through definitive policy and the operations analysis essential to the administration of that policy, system design specifications may be drawn up upon which to base detailed design of the human-service subsystems.

NOTES TO CHAPTER 8

1. R. Sadacca and R. Root, "A Method of Evaluating Large Numbers of System Alternatives," *Human Factors Journal* 10 (February 1968) :5–10; W. Knowles et al., "Models, Measures and Judgments in System Design," *Human Factors Journal* 11 (December 1969) :577–90; H. Puscheck and J. Greene, "Sequential Decision Making in a Conflict Environment," *Human Factors Journal* 14 (December 1972) :561–72.

2. W. Wheaton, "Operations Research for Metropolitan Planning," *American Institute of Planners Journal*, November 1965, pp. 250–59; G. Barnhart, "Social Design and Operations Research," *Public Health Reports* 85 (March 1970) :247–50; M. Springer, "Social Indicators, Reports, and Accounts: Toward the Management of Society," *The Annals of the American Academy of Political and Social Science* 388 (March 1970) :1–13.

3. J. Burgess, "A Goals Versus Process Orientation Policy in Community Mental Health," *Social Psychiatry* 10 (1975) :9–13.

4. H. Halpert et al., *The Administrator's Handbook on the Application of Operations Research to the Administration of Mental Health Systems* (National Clearinghouse for Mental Health, National Institute of Mental Health, Washington, D.C., 1970) ; G. Murray and L. Klainer, "On Making Systems Analysis Results Operational in Comprehensive Health Planning," *American Journal of Public Health* 62 (July 1972) :980–84.

5. E. Bursk and J. Chapman, *New Decision-Making Tools for Managers* (New York: The New American Library, Inc., 1965) ; H. Fox, "Toward an Understanding of Operations Research Concepts," *Management Services*, July 1970, pp. 23–36.

6. D. Carson, "Human Factors and Elements of Urban Housing," in *Human Factors Applications in Urban Development* (Riverside Research Institute, New York, 1970) .

7. R. Buel, "Alternatives to the Automobile," in *Dead End: The Automobile in Mass Transportation* (Englewood Cliffs, N.J.: Prentice-Hall, Inc., 1972) pp. 157–211; "Transportation" *Science News*, 104, (1973) :170 f.

8. R. Archibald, "Introducing Technological Change in the New York City Fire Department," in *Human Factors Applications in Urban Development* (Riverside Research Institute, 80 West End Avenue, New York, 1970) ; E. Krendel, "Social Indicators and Urban Systems Dynamics," in *Human Factors Applications in Urban Development* (Riverside Research Institute, New York, 1970) ; "Science Awakens to America's Fires," *Science News* 104, no. 22 (December 1, 1973) :337–52.

9. M. Gold and J. Williams, "The National Survey of Youth—Search of Official Records," Research Center for Group Dynamics, Institute of Social Research, University of Michigan (Ann Arbor, Mich., 1969) ; J. Williams et al., "The Incidence of Detected and Undetected Delinquency in the United States from 1964 to 1967: A Benchmark Study," Research Center for Group Dynamics, Institute

for Social Research, University of Michigan (Ann Arbor, Mich., 1970) ; H. James, "Probation: An Alternative to Prison," in *Children in Trouble* (New York: David McKay Company, Inc., 1970) , pp. 88–104; *National Advisory Commission on Criminal Justice Standards and Goals*, Interim Report (Washington, D.C., 1973) .

10. K. White, "Primary Medical Care of Families—Organization and Evaluation," *The New England Journal of Medicine* 277 (October 19, 1967) :847–52; B. King and E. Sox, "An Emergency Medical Service System-Analysis of Workload," *Public Health Reports* 82 (November 1967) :995–1008; O. Anderson, "Rising Costs Are Inherent in Modern Health Care Systems," *Hospitals* 43 (February 1969) :50–52; "The Health Care Crisis," *Hospitals* 43 (September 1969) :87–98; "Early Malnutrition and Human Development," *Children* 16 (November 1969) :211–15; H. Ennes, "The Insurance Industry Looks Ahead at Health Care in the 1970's," *Public Health Reports* 85 (February 1970) :117–22; M. Tucker, "Effect of Heavy Medical Expenditure on Low Income Families," *Public Health Reports* 85 (May 1970) :419–25; S. Wade, "Trends in Public Knowledge about Health and Illness," *American Journal of Public Health* 60 (March 1970) :485–91.

11. W. Horvath, "The Systems Approach to the National Health Problem," *Management Science* 12 (June 1966) :391–95; M. Blumberg, "Systems Analysis and Health Manpower," *Journal of the American Medical Association* 201 (September 1967) : 178–79; H. Adelman, "System Analysis and Planning for Public Health Care," *Archives of Environmental Health* 16 (February 1968) :258–63; A. Bennett, "Systems Engineering," *Hospitals* 43 (April 1969) :171–74; P. Rogatz, "Planning the Health Care System," *Hospitals* 16 (April 1970) :47–50; S. Garfield, "The Delivery of Medical Care," *Scientific American* 222 (April 1970) ; C. Flagle, "The Role of Simulation in the Health Services," *American Journal of Public Health* 60 (December 1970) :2386–94; M. Pilot, "Health Manpower 1980," *Training and Development* 14 (Winter 1970) :2–3; E. Weinerman, "Research on Comparative Health Service Systems," *Medical Care* 9 (May 1971) :272–89.

12. D. Vail et al., "The Relationship Between Socioeconomic Variables and Major Mental Illness in the Counties of a Midwestern State," *Community Mental Health Journal* 2 (Fall 1966) :211–12; L. Lippman, "Deviancy: A Different Look," *Mental Retardation*, June 1970, pp. 6–8; R. Nader, "Task Force on the Community Mental Health Centers Program," Center for the Study of Responsive Law (Box 19369 Washington, D.C., 20036, 1972) ; H. Schulberg, "The Mental Hospital in the Era of Human Services," *Hospital and Community Psychiatry* 24 (July 1973) : 467–72; S. Southard, "The Process of Planning for Program Reorganization," Professional Services and Training, Georgia Mental Health Institute (Atlanta, Georgia, 1973) .

13. L. Goodwin, "Poor People and Public Policy," pamphlet (Brookings Institution, Washington, D.C., 1973) ; R. Morris, "Welfare Reform 1973: The Social Services Dimension" *American Psychologist*, August 1973, pp. 515–22.

14. A. Packer, "Applying Cost-Effectiveness Concepts to the Community Health System," *Bulletin of the Operations Research Society of America* 16 (March 1968): 227–53.

15. M. Herz et al., "Day Versus Inpatient Hospitalization: A Controlled Study," *American Journal of Psychiatry* 127 (April 1971) :1371–82; R. Taylor and E. Torrey, "Mental Health Coverage Under a Major Health Insurance Plan," (National Institute of Mental Health, 5600 Fishers Lane, Rockville, Maryland, 20852, 1972) .

16. J. Christensen and R. Mills, "What Does the Operator Do in Complex Systems?" *Human Factors Journal* 9 (August 1967) :329–40.

9

Developing
Systems Design Specifications

A documented approach to the design of human-service subsystems requires first a clear-cut statement of design objectives. To avoid equivocation, these should be quantified, with a tolerance or latitude of error specified, e.g., to increase the average educational grade-level achievement of young poverty-group adults from 9.1 to 11.5 $\pm$ .3 grades in a five-year program.

Given the ultimate objectives of the specific project or program within the subsystem, specification of design requirements may then proceed on the basis of a documented format. The format may then serve as a comprehensive engineering basis for design.

Various formats may be employed, but general areas of interest may best include the following comprehensive substantive content:

1. Target population or public sector for whom services are to be provided. Hard data or estimates of volumes of population served or incidents and prevalence of needs should be included.
2. Operational requirements describing the current conditions of services, and salient areas for redesign or new design to accomplish subsystem objectives. Descriptions of problem areas and constraints are also provided.
3. System or subsystem description with detailed specification of the human-service purpose (s) being served. This should include a narrative and graphic summary of the anticipated method of operation and support of the system, to include general facilities, resources, manpower and special skill requirements, manual and machine-processed control or record keeping, funding sources and legislative advocates. Assumptions and special problem areas are also identified.
4. Operations and maintenance summary to include operational sequences through initial developments to routine operations

152

and continuing support. Classes of personnel involved, team performance, periods of service action, lists of vehicular and other equipment requirements are included.

5. Position descriptions including duties and tasks selectively combined with multiple time-shared and sequential activities essential to the development, operation, maintenance and control of the subsystem. Available skills and special training requirements are included.

6. Operating environment, physical, social and biological conditions of performance, e.g., toxicity in fumes; outdoor, indoor temperature; noise and lighting conditions; office, laboratory, ghetto settings; conditions of fatigue, malnourishment of clients, special hazards, etc.

7. Manning estimates to include a statement of the number and kinds of personnel required, the number of shifts required by program area in the subsystem. Criteria and special conditions are presented in sufficient detail to justify the estimates.

8. Provisions for evaluation to include all evidence associated with the accomplishment of overall system or subsystem objectives, and costing and accounting provisions to serve as the basis for cost-effectiveness assessment.

9. Provisions for developmental and operational administrative review, to include progress reporting periods, audio-visual methods of presentation, and administrative control procedures to be employed.

System design requirement specifications may serve a central function in directing operational design and control. Such carefully prepared "specs" may then be used as the master plan for implementation and operation. Design and operation may be completely controlled and implemented by a single administrative body such as a coded department, tax funded and operating under the auspices of the state or federal government; or, as is generally the case with aerospace systems, design responsibility may be let out to a prime contractor and his subcontractors, controlled through detailed specifications with extensive contract monitoring.

SPECIFICATION EFFECTIVENESS

The use and effectiveness of system design specifications in the human-service industry has not been extensively demonstrated, particularly where definitive outcome design objectives are involved.

Largely on an a priori basis, however, and based on the success of aerospace projects employing the technology, it might be safely assumed that such design control is both appropriate and feasible in human services.

The feasibility and effectiveness of such specification control measures in human services has to some extent already been demonstrated for mental health operations in Illinois. Early work in community mental health was predicated in part upon military psychiatric experience which indicated that chronicity of mental illness in soldiers was spawned by institutionalization. Evacuation to rear echelons and sophisticated psychiatric diagnostic labeling from simple battle fatigue resulted in high rates of chronic Army and V.A. hospital tenure, and depleted the combat manpower reserves. World War II experience indicated that maintaining expectancy of return to the front line for the overwrought or dazed G.I. combatant, and keeping him nearby resulted in high rates of recovery.[1] Thus, the community mental health movement was designed to minimize removal of mentally ill patients to third echelon or state hospital facilities while maintaining their close contact with the community. Following the 1963 Community Mental Health Act (Public Law 88–164), however, the new mental health centers were not doing this. They seemed rather to envision their charge as one of finding and gathering severely disturbed persons to place in state hospitals. The more funding they received, the more cases they found for the state hospital (see Figure 17). This was contrary to the intent of the Act, and the purpose for which grant money was being awarded.[2]

In October of 1970, the Illinois Department of Mental Health issued guidelines specifically instructing grant-supported mental health clinics to serve clients in the community. Excessive extrusion rates or admissions to state facilities, it was iterated, would be grounds for reduction of grant funding. Figure 17 shows the impact of such guidelines or program design specifications on subsequent admission rates (after 1970) from east-central Illinois. This may be interpreted to mean that the specifications had a direct bearing on the mental health clinics program design effort while, before, only random or conflictual policy control was operating. Such specifications must of course be monitored and enforced, but the potential ramifications in preparing and employing design specifications for human services harbinger well for future service developments.

Another area in which design specifications were functionally employed was in the delivery of comprehensive community services for the adult mentally retarded population in Macon County, Illinois.

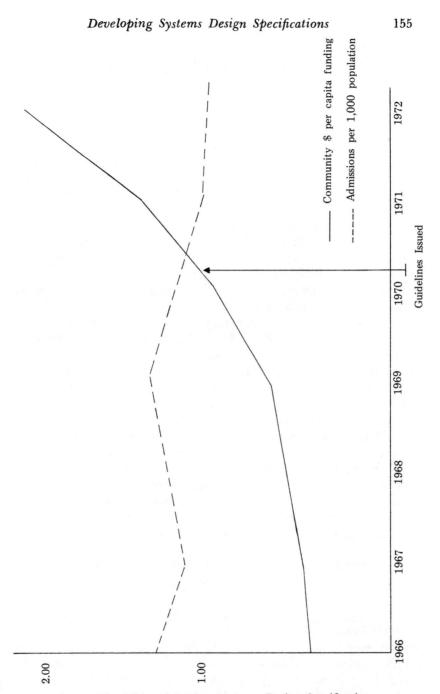

Fig. 17. The Effect of Issuing Program Design Specifications
on the Admission Rates to Mental Hospitals

In September of 1970, a "Request for Proposal" (R-for-P) was issued to several community service agencies soliciting proposed program descriptions for such services. The R-for-P essentially provided a systems design specification, requiring that a complete comprehensive and effective gamut of services be provided a specified adult mentally retarded population during the first and subsequent years, viz., screening and placement, medical services, counseling, financial and legal support, transportation, referral, vocational and workshop training and placement, etc.[3] Service strategies, manpower, facilities, equipment, budgeting and costing data were required in the proposal response.

The careful development and preparation of such a system design specification served a number of purposes in advancing this particular human-service subsystem component. It served to provide more or less precisely structured control of the program design elements essential for such a mentally retarded population. A detailed operational definition of the target population minimized equivocation for whom, and of what severity in mental deficiency, the services were to be provided, as well as where the populations were to be found. Moreover, it served an essential didactic function for the bidders. In drawing upon the sophisticated expertise of state-employed personnel for substantive specification content, bidders were made aware of the state-of-the-art in services for the mentally retarded. The R-for-P did, in fact, result in a comprehensive proposal that was subsequently funded, and has since been revised and updated for each funding period to address current needs of the mentally retarded in Macon County. The effective employment of such a design specification for the mentally retarded provides further evidence that system design specfications are functional. Their potential for more widespread employment in the varied and sundry operations of other human-service industries must therefore be seriously considered for future progress in human services.

DEVELOPMENTAL AREAS FOR HUMAN-SERVICE SYSTEM DESIGN SPECIFICATIONS

The broad systems implication in design of human-service subsystems using a design specification format is that each area or subsystem may become structured with some measure of design control. A glance at several of the subsystems and their gross requirements may be helpful in setting a perspective for preparation of such design specifications.

The domiciliary and architectural subsystem, for example, may involve a goal setting of slum clearance with the relocation of low-income housing, e.g., to have 50 percent of all low-income families in a specified dilapidated housing area relocated in functional housing within a five-year period. This, as a major design project, might entail the preparation of general and detailed control specifications, identifying the target population, the operational housing requirements, and a total system description of the post-five-year period housing complex. Development phases, construction, and maintenance summaries would be included, together with the manpower and skills required to operate the housing complex. Detailed design considerations, with specified physical, social, and behavioral criteria, would be included.[4]

A surface traffic subsystem specification may be developed about such hypothetical goal formulations as reducing the required personal travel time from zoned urban fringe areas at peak travel hours by 25 percent within a three-year development program. Specifications may then be prepared on the basis of optimally-OR determined necessary system components of the transport system.[5] Detailed specifications are then largely developed about the detailed design aspects of proposed systems determined most optimally suited to meet the travel design objectives (refer, for example, to the Ontario Department of Transportation and Communications design documents for the electromagnetic mass-transit system design in Toronto, Canada).

Air traffic subsystems may also be aggressively advanced through the use of carefully prepared design specifications. For example, the problematic conditions of inordinate workload pressures and the attendant marginal safety of ground controller functions may be structured in a design specification about a reduction of measured controller workload per unit of time in peak arrival-departure activity. Several studies pertinent to such a design specification have already begun to take shape in detailed perceptual-motor tasks, motivational and work-aid requirements.[6]

Preparation and implementation of design specifications are also feasible in fire protection subsystems. For example, in filtering through the complex of high false-alarm rates, vacant building fires, hostile ghetto environments, etc., a more inclusive design specification goal may be set, such as: "Reduce overall injury and fatality rates and property damage attributable to uncontrolled burning by 30 percent within specified urban boundaries over a five-year period." With such a design parameter, about which to orient and organize design efforts, the total fire-fighting technology, and subsidiary areas of control such as harassment, may be brought to bear to accomplish major and minor design objectives. System design specifications may serve in controlling the design approaches and detailed areas for integration about the design goals and performance requirements.[7]

In the criminal justice subsystem, documented design specifications might be written about a design goal of recidivism. A goal statement might be formulated, such as, "Reduce the rate of multiple offenders by 25 percent within a specified correction system during a three-year period."

The design specifications would then necessarily involve a full gamut of design considerations to be integrated about the design goal of reduced recidivism.[8]

A sanitation design requirements specification might involve improved recycling operations, e.g., "within the next decade to increase derived useful products from liquid, solid and gaseous wastes to 175 percent of the 1974 useful products level." As a specified design goal, about which programs and development projects might be integrated, detailed design specifications can be prepared by appropriate technical experts to direct government and industrial officials into essential and subsidiary development areas.[9]

A health-care subsystem design specification may revolve about such goals as improved services in preventive medicine to identified lower socioeconomic population sectors, addressing epidemiological problems, infant mortality, life expectancy, etc. For example, a goal statement may consist of the reduction of extant, reported and unreported, venereal disease rates per 1,000, age 18-to-30 population by 50 percent within a specified geographic area over a two-year period. Such a project, written up in detailed design requirements specifications, would direct a total complex of activities, employing current state-of-the-art in medicine, communication, microbiology, social service, etc., to be integrated about the definitive objective of V.D. reduction.[10]

The successful use of design specifications for mental health subsystem components has been described earlier in this chapter. Future design work in this area must be predicated on the growing conviction among mental health professionals that a patient's problems are as much rooted in a community's disintegrated condition and its fragmented care-giving system as they are in the patient's psyche, per se.[11] Design objectives may be variously stated within this context, but, to be meaningful in design specification development, must be properly circumscribed with population and geographic parameters. Thus, a statement such as follows may serve in specification development:

"To restore to normal vocational or domestic functioning, 80 percent of all first-time patients, either petitioned or themselves applying for admission as inpatients to mental hospitals, within a specified community by the end of a three year period."

Such a specifically stated goal may then permit the development of detailed design specifications, integrating all known and developmental contingencies bearing upon the established needs of this population to accomplish objectives.[12]

The broad spectrum of a public welfare subsystem likewise must be treated by specific components to permit workable systems design specifications to be developed. A basic goal orientation in keeping with welfare policy, for example, migh* be stated something as follows: "Within a designated geographic area, increase the rate of job placement of welfare mothers heading household by 30 percent and fathers heading household by 50 percent after a five-year period."

Design specifications may then be prepared in detail to include provisions for all constraints and limitations, such as child care, lack of nutritional vitality for work, skill and educational deficiencies, failure of the work ethic, opportunity limitations, etc.[13]

LIMITATIONS, CONSTRAINTS AND CAUTIONARY NOTES IN THE PREPARATION OF SYSTEM DESIGN SPECIFICATIONS

The straightforward descriptive format outlined earlier in the chapter included such fundamentally identifiable system parameters as target population; conceptualized systems of service; operations and maintenance summary; required job positions; the physical, social and biological environment of the system; and the manning, training, evaluation and managerial requirements of the system. However, a number of more subtle system requirements and format limitations should be noted:

1. *System design objectives.* In human-service systems the way in which operational objectives are defined could spell the success or failure of the system. They should be neither so broad and general as to be vague and indeterminate nor so specific and restrictive as to be unrealistic or to balk creative design solutions for service delivery.[14]

2. *Target population.* Careful study is necessary at the outset to define the specific needs to be served by volume and outcome expected.

3. *Validity of service.* Operations research studies should largely have determined that the system design requirements spelled out in the specification are valid, i.e., will reduce traffic congestion, will rehabilitate mental patients, etc. However, design specifications must assure that the operational system will perform in such a way as to do what is intended.

4. *Service effectiveness.* The specification must make clear the necessity for a clear-cut evaluation procedure and routine, and that criteria employed are directly related to desired outcome.

5. *Administrative control.* The specification should also emphatically require assurance of effective administrative skills to be employed in project coordination. This has often proved to be a major cause for design failure, cost overruns or operational sterility in both military and human service systems.[15]

6. *Political control and support.* Design specifications in a human-service context must be keenly sensitive to public biases, prejudices and the general emotional climate spawned by current crises in energy, fuel, population, pollution, inflation, etc. In the early seventies, for example, the Supersonic Transport was scrapped for want of public support or support of the majority spokesmen who were not convinced of its functional value.

7. *Design for adaptability.* The changing milieu of technology and rapidly changing needs of society must also be given consideration in the preparation of design specifications. Military systems have often cost in the neighborhood of a billion dollars to develop, only to be outmoded and become useless within a few years after attaining operational status. Human-service systems, if adaptable to changing needs, constraints and limitations, could avert such wasteful extravagance from poor planning and lack of adaptability.

The development of system design specifications for definitive technical direction of human-service systems appears to be entirely feasible. Their realistic value, however, must lie in the erudition with which they are prepared. Such specifications cannot be issued on a slipshod, general and vague basis if effective concrete design is to result. They must above all be based on the highest level of technology available in the subsystem component being addressed, and, to be realistically based, due consideration must be given to the ways and means to achieve public support, and to costing and manpower constraints.

NOTES TO CHAPTER 9

1. M. Parrish [now with the Illinois Department of Mental Health] "Some Concepts of Military Psychiatry," in *Military Psychiatry Consultation and Group Therapy*, 1967; A. Glass, "Military Psychiatry and Changing Systems of Mental Health Care," *Journal of Psychiatric Research* 8 (1971):499–512.

2. R. Nader, Task Force on the Community Mental Health Centers Program (Center for the Study of Responsive Law, Box 19367, Washington, D.C. 20036, 1972).

3. Adolf Meyer Center Request for Proposal for the Development and Delivery of Comprehensive Community Services for the Adult Mentally Retarded (over 18). Macon County, September 1, 1970. Illinois Department of Mental Health, Zone VI, Subzone 2, Decatur, Illinois.

4. R. Smith, "Job Analysis as a Basis for House Arrangement," *Journal of Home Economics* 54 (1962):458–59; R. Mitchell, "Some Social Implications of High Density Housing," *American Sociological Review* 36 (February 1971):18–29; S. Levin, "Human Engineering in the City," in *Proceedings of the Sixteenth Annual Meeting of the Human Factors Society* (Santa Monica, Calif., 1972), pp. 302–06; J. Siegel, "The Role of Industry in Training the Hard Core Unemployed," in *Human Factors in Urban Development* (Riverside Research Institute, 80 West End Avenue, New York, 1970); A. Chambers and I. Seymour, "Improving the Management of Public Housing Authorities: The Role of Human Factors," in *Proceedings of the Seventeenth Annual Meeting of the Human Factors Society* (Santa Monica, Calif., 1973), pp. 352–58.

5. F. Rossini and T. Tanner, "Transportation, Communication and Population Distribution," in *Proceedings of the Sixteenth Annual Meeting of the Human Factors Society* (Santa Monica, Calif., 1972), pp. 301–06; L. Hoag and S. Adams, "Human Factors in Urban Transportation," in *Proceedings of the Seventeenth Annual Meeting of the Human Factors Society* (Santa Monica, Calif., 1973), pp. 126–35.

6. V. Hopkins, "Human Factors in the Ground Control of Aircraft NATO AGARDograph," 1970; L. Laveson and C. Silver, "The Employment of a Spoken Language Computer Applied to an Air Traffic Control Task," in *Proceedings of the Sixteenth Annual Meeting of the Human Factors Society*, (Santa Monica, Calif., 1972), pp. 410–15; K. Teel, "Human Resources Management: A Case for Involving Employees in Job Design," in ibid. pp. 465–66; J. Miller, "Visual Behavior Changes of Student Pilots Flying Instrument Approaches," in *Proceedings of the Seventeenth Annual Meeting of the Human Factors Society* (Santa Monica, Calif., 1973), pp. 208–14.

7. R. Archibald, "Introducing Technological Change in the New York City Fire Department," in *Symposium Proceedings, Human Factors Applications in Urban Development*, (Riverside Research Institute, 80 West End Avenue, New York, 1970); W. Teichner and D. Olson, "A Preliminary Theory of the Effects of Task and Environmental Factors on Human Performance," *Human Factors Journal* 13 (August 1971):295–344.

8. M. Gold, "Juvenile Delinquency as a Symptom of Alienation," *Journal of Social Issues* 25, no. 2 (1969):121–35; L. Mosher, "Why Prisoners Riot—And Reform Fails," *National Observer*, August 31, 1970; *Division of Research and Demonstrations*, "Delinquents Don't Like Themselves—That's Partly Why They're Delin-

quent," Research Utilization Branch Office of Research, Demonstrations, and Training, Social and Rehabilitation Service, Department of Health, Education and Welfare, Washington, D.C., Volume III, Number 7, March 1, 1970; L. Barber, "Changes in Self-Concept Among Delinquent Boys in a Therapeutic Community," *Dissertation Abstracts* 33 (January 1973) :3331–B.

9. E. Steel, *Water Supply and Sewerage* (New York: McGraw-Hill Book Company, 1960) ; V. Ehlers, *Municipal and Rural Sanitation* (New York: McGraw-Hill Book Company, 1965) ; "Sewage into Shellfish," *Newsweek,* December 24, 1973, pp 101f; "Garbage Gas Is Put to Work," Associated Press, Palos Verdes, Calif., December 24, 1973.

10. E. Gardner and J. Snipe, "Toward the Coordination and integration of Personal Health Services," *American Journal of Public Health* 60 (November 1970) : 2068–78; P. Tobias et al., "Human Factors in the Design of a Computerized System for a Neighborhood Health Clinic," in *Proceedings of the Sixteenth Annual Meeting of the Human Factors Society* (Santa Monica, Calif., 1972) , p. 247; M. Rudov, "Consumer Health Education: An Analysis," in *Proceedings of the Seventeenth Annual Meeting of the Human Factors Society* (Santa Monica, Calif., 1973) , pp. 204–07; J. Stanfield, "Some Human Factors Aspects of Emergency Medical Care," in ibid., pp. 200–03.

11. H. Schulberg, "The Mental Hospital in the Era of Human Services," *Hospital & Community Psychiatry* 24 (July 1973) :467–72.

12. A. Hollingshead and F. Redlich, *Social Class and Mental Illness* (New York: John Wiley and Sons, Inc., 1964) ; P. McCullough, "Mental Health Programs and Boundaries," in *Systems Approach to Program Evaluation in Mental Health* (Western Interstate Commission for Higher Education, Post Office Drawer "p", Boulder, Colorado, 1970) , pp. 47–63.

13. A. Hovater et al., "Industrial Engineering in Vocational Rehabilitation," *Rehabilitation Literature* 30 (November 1969) :322–25; E. Corlett, "Efficient Labor Utilization in a Developing Economy," *Human Factors Journal* 12 (October 1970) : 499–502; C. DeBow, "Job Task Requirements Analysis in Career Progression Systems," in *Proceedings of the Sixteenth Annual Meeting of the Human Factors Society* (Santa Monica, Calif., 1972) , p. 473.

14. J. Lyman, "Measuring Metafacts," *Human Factors Journal* 11 (February 1969) :3–8.

15. J. Burgess, "Who Has the Administrative Skills in Mental Health?" *Public Administration Review,* April/March 1974, pp. 164–66.

10

Task and Training Functions
Analysis in Human-Service Systems

The now classical human-factors systems approach to task and training functions design is predicated upon careful analyses of system and subsystem performance requirements. The early work of Robert Miller at the American Institute for Research (1953) set forth a clearly defined human-factors methodology for the analysis of tasks in evaluating man-machine relations, and for determining performance skills and training requirements. Such data requirements were later incorporated into military design specifications (U.S. Air Force Military Specification MIL-D-26239A, 1961), which continue to govern design requirements for the personnel subsystem in military operations.

Since this early work, increasingly sophisticated developments have continued in the analysis of human operator skills and system training requirements. Task assignment problems have been addressed in an attempt to optimize task elements for which the human operator is best suited,[1] while the practical problems of limited available skills and training time during a period of military enlistment have also been considered.[2] Multiple training simulator and tasks skill acquisition studies have also been completed to determine the most efficient ways in which system management and performance skills may be acquired by the human operator.[3]

However, again, in the human-services sector such methodological developments are minimal or virtually nonexistent. In architectural design, Sommer[4] laments the fact that only infrequently are functional behavioral questions asked concerning the design of classrooms, dormitories, parks, prisons, etc. When questions are asked, they are posited in the context of a predetermined functional base: What is the size and shape of prison cells preferred by guards and prisoners—the best lighting, optimal distance between bars, colors most suitable for prison walls, etc.? Such questions, even though rarely asked, ignore the more basic question about how the prisons should relate to society's goals of rehabilitation. This seems to parallel the total human-service system concern—how can we train more police more efficiently, attract more young men to the priesthood, obtain increased funding for the training of physicians, motivate more welfare recipients to obtain productive employment, etc.? Such questions, in keeping with Sommer's lament, are more bent on perpetuating the stereotyped, fixed functions currently operating than on the accomplishment of society's missions.

Approaches to task and training functions analyses must again be predicated on, and oriented to, desired system outcomes. Meanwhile, only subsidiary questions on tasks, training and manpower may be addressed.

The present chapter attempts to relate the state-of-the-art in human-factors training and manpower analytic methods to human-service industries. Brief review is made of training and manpower dilemmas in several human-service subsystems. The mental health subsystem is addressed in greatest detail, since this area was studied in some depth by the author. A specific application of human-factors personnel and training analytical methods to the latter subsystem is also described.

CURRENT HUMAN-SERVICE TRAINING DILEMMAS

A major disparity between available skills and the identification of training requirements in human services often occurs due to lack of clear-cut service objectives. In police work, for example, analyses are frequently made on the basis of currently specified patrol duties.[5] Task and skill requirements are thus not generally determined from data generated in a prevention and control mission, as in the analysis of delinquency completed by *Space-General Corporation* (1965).[6] Manpower requirements in the sanitation field likewise appear to be based on gross estimates and general, if not vague, projected societal needs.

In the health-care field, the burgeoning need for medical specialists often occurs on a status quo basis, e.g., medical doctors refuse to accept new patients when their patient load places excess demands on their time. Here the manpower solution is one of a foregone conclusion— simply that more doctors or more equitable distribution of their numbers are needed, rather than of what specific tasks and skills are required for training.[7] The status quo of the medical system, however, is increasingly being called into question on a practical basis,[8] while others are calling for more detailed studies to establish health-care system manpower needs.[9]

Manpower requirements in the mental health subsystem are also largely predicated on professional status rather than on task and skill requirements analysis. Psychiatric specialization is now seen as of limited value in working with high-risk populations.[10] Even psychologists are severely called to task on their contribution. An entire issue of the *American Psychologist* was devoted to questions concerning the education and utilization of psychologists.[11]

Training requirements need also to be established in the welfare subsystem, as with counselors and trainers, who may be sorely unsuited to their tasks, or misoriented to the essential needs of the poor.[12]

Indeed, the current human-service training dilemma is but a part of the more major systems design problem. Knowles and others[13] have seen the way in which the personnel subsystem relates to a system as fundamentally important to total design. From a systems standpoint, many functions and tasks are interchangeable among human operators given nominal training and experience. The critical tasks in a system are those that presumably require special skills, aptitude or training, which often introduce inordinately high costs and formidable, if not impossible, assessment problems. Moreover, the system manpower needs are often confounded by vested credentialism and personal ego involvement by designers, which, though assuredly operating in military systems design, assume major proportions in human-service industries.[14]

THE MANPOWER MUDDLE IN THE MENTAL HEALTH SUBSYSTEM

In the rapidly changing milieu of mental health services, a multiplicity of solutions are frequently offered to the ostensible problem of manpower shortages. These may range from advocating the employment of college graduates to perform therapy, or natural neighbors and specially trained paraprofessionals,[15] to the revivification of demands for new but highly trained professionals.[16] Others have proposed the training of middle-level personnel for therapy,[17] or generalists for community services.[18] Several writers have come to see the new mental health vogue as one of social action and community programs for which new skills, training and special talents are required.[19] Vidaver[20] holds that high-level professionals in mental health are a thing of the past, and that we must develop alternative manpower resources.

In the confusion of current mental health services, only rarely is a kind of professional heresy proposed that an analysis of requisite skills be made to demonstrate that many job functions can be done without high-level graduate training.[21]

Ideally, from a human-factors standpoint, a complete and detailed task analysis should be made to determine skills and training requirements within the framework of a total mental health mission. Total mission sequences for mental health patients are currently generally not indicated, while traditional therapeutic and incipient rehabilitative services continue as the chief mode of operation. At the Adolf Meyer Center in Decatur, Illinois, an interim human-factors technique was developed for the management of current personnel mental health functions. This was modeled, to some extent, after a human-factors systems analytical approach, and, though the skill patterns are still tied to conventional modes of operation, it represents an early inroad for task and training functions analysis in human services.[22]

A TECHNIQUE FOR PERSONNEL MANAGEMENT AND TRAINING IN MENTAL HEALTH

Over the past two decades, and particularly since enactment of the Community Mental Health Act of 1963, the rapidly changing milieu of mental health practices increasingly imposed new and often unproved skill demands on the mental health professions. In 1971, a "Generalist Series" of job classifications was issued in the Illinois Department of Mental Health to provide manpower in meeting the state's changing mental health needs.[23] Each of the state's seven administrative regions was given responisbility for preparing its employees through training program design at various kinds and levels of skill requirements. Training programs were to be designed about the work role of the employee's required performance, with development responsibility to be shared by training administrators and employees' supervisors.

In implementing the generalist training series for an east-central region of Illinois,[24] the development model employed was that of a military systems approach to training.[25] The theory of analysis was simply that a detailed description of work requirements should match the educational and/or experiential background of the employee. At the outset, only descriptive work areas were employed, with more sophisticated analyses of work elements to be determined as more comprehensive data became available on mental health service functions.[26]

MENTAL HEALTH FUNCTIONAL JOB REQUIREMENTS

Each of 70 employees in a generalist series of job classifications completed a detailed job description of the work he had been doing over a period of several months. Immediate supervisors reviewed work descriptions, and returned them to the employee for revision or discussion. The job analysis format included the following functions described in detail by each employee:

Administration. (Functions were described in as much detail as possible in such areas as program planning, budgeting, personnel planning, etc., together with the frequency required, etc.)

Supervision of Staff. (Functions, numbers, kinds and levels of staff, etc.)

Supervision of Residents. (Activities, numbers, and kinds of residents, etc.)

Resident Discipline. (Interaction with residents and discipline modes, verbal and physical restraints required, etc.)

Housekeeping. (Cleaning and other maintenance chores within the unit, requirements for resident participation, etc.)

Diagnosis and Rehabilitation Planning. (Kinds of diagnoses made and programs planned for residents, clinical responsibility, number of concurrent cases, etc.)

Psychotherapy. (Dyadic therapeutic techniques employed, numbers of cases, frequency of sessions, etc.)

Family Counseling. (Nature of sessions, location, whether conjoint, frequency per case, etc.)

Intrastaff Coordination. (Required interactions with other staff on patient care for continuity, etc.)

Interagency Coordination. (Interactions with other agencies for case consultation, names of agencies, objectives, etc.)

Community Organization. (Interactions and kinds of organizational development objectives, responsibilities, etc.)

Group Therapy. (Group therapy sessions and techniques employed, size of groups, frequency of sessions, etc.)

Behavior Modification. (Procedures employed, requirements for timing, appropriateness and effectiveness of reinforcement, etc.)

Activity Therapy. (Specific kinds of activities such as occupational therapy, games and other forms of recreation, bibliotherapy, etc., and number and frequency of sessions, etc.)

Rehabilitation Counseling. (Specific kinds of counseling, such as budgeting and other domestic skills, prevocational and job interview, number and frequency of sessions, etc.)

Teaching/Training. (Kinds and content, whether tutorial, for staff or residents, size and frequency of classes, etc.)

General Health Applications. (Frequency and kinds of medication or first aid applied, etc.)

Evaluation. (Nature of evaluation, whether individual or program, number of subjects and frequency of studies, etc.)

Research. (Types of research, levels of responsibility, controls applied, statistical methods, etc.)

Report Writing. (Content, level of interpretation, objectives, readership, etc.)

Public Relations. (Nature of contact with public sector, relationship to programs, frequency of contacts, etc.)

For new or unfilled jobs, supervisors completed the functional job descriptions themselves based on the conceptual framework for which the job was designed.

EMPLOYEE'S EDUCATION AND SKILLS INVENTORY

Employees then completed a description of their training and experience in those areas corresponding to the functional job description, including places and dates of course work or job experience. These, too, were reviewed by the supervisor for discussion and clarification with the employee.

JOB SKILLS AND TRAINING REQUIREMENTS PROFILE

Training requirements profiles were then constructed, employing preliminary scoring procedures, as illustrated in Figure 18. A task or job requirement indicated to be a primary responsibility in the position was assigned a score of 100; or, on a judgmental basis, some proportion of 100 was assigned the task depending upon the scope and extent of responsibility indicated.

In like manner, education and experience were scored relative to a sum of 100. Education was thus allowed up to a maximum score of 50 for any given function, as was experience. The sum of these (solid line) was then plotted against the functional job requirement (broken line) to obtain a measure of specific skills deficiency in the job. Gaps marked "x" in Figure 18 present the areas of deficiency.

NOTE: Solid line indicates training and experience.
        Broken line presents job requirements.
        Outside gaps (x) indicate skills and training requirements.

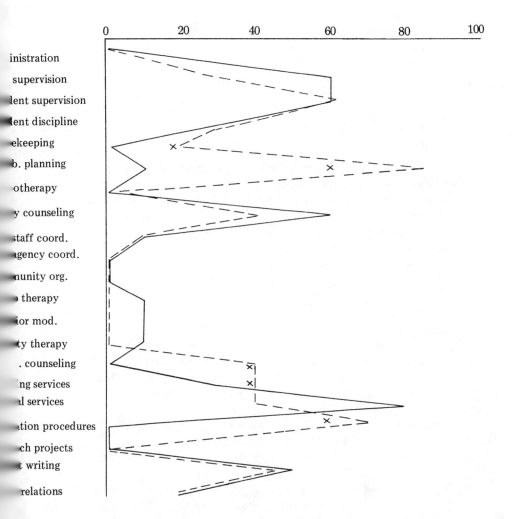

Fig. 18. Job-Skills and Training-Requirements Profile. From D. Nellis and J. Burgess, "An Administrative Technique for Personnel Management and training in Mental Health," *Public Administration Review,* September–October 1974, p. 498.

The availability of such a profile provides for a number of uses and advantages in training and personnel administration:

—Basic training program design is based on a tailored-training requirements inventory
—A communication tool is provided in clarifying job elements between supervisor and employee
—It may be used as a precise monitoring device to update skill requirements and required training functions
—Serves as a conglomerate standard for certifying employees at given levels of specified generalist positions
—Used in assessing skills for transferring employees to new positions
—Review process sensitizes employees and their supervisor to the various nuances and skills required in working with mental patients, e.g., in serving as an advocate in interagency coordination, etc.

The method thus permits tailoring of training functions to specific work-oriented skill requirements. Areas of skill and information deficiency may be readily determined, with training directly bearing on the employee's work performance. This is in contrast to training programs administered by gross nomenclatures, often irrelevant to the employee's job requirements. Such course training as "encounter therapy" or "transactional analysis," for example, may be both costly and of little practical value, having little or no bearing on his functional job requirements.

Although generally lacking definitive goal direction, continuing innovative programming in mental health will call for increasing variation and diversity of personnel functions, frequently deviating from the conventional mental health professional skills. This will necessitate the use of such human-factors analytical methods in personnel administration and training program design.

TRAINING FOR ADMINISTRATION OF HUMAN-SERVICES INNOVATIVE PROGRAMMING

Management of human-service system development programs would currently appear to be a most important but severely neglected skill area in light of the need for new and progressive systems in virtually all human-service subsystems. Effective management of innovation appears to be the key to advances in all areas of human service technology.

Recent findings have shown that innovation will not likely be efficiently achieved through such basic research as in the biochemistry of schizophrenia, socioeconomic factors in delinquency, etc., but from missions-oriented research that is goal directed and programmatically based.[27] Nor will innovation suddenly emerge when all the facts are in; rather, it will occur only through the orchestrating of well-planned and well-managed innovative human-service systems design. Such systems may be variegated, diverse and individuated, but they must indeed be managed with all the administrative and technological skills available.

CORE ADMINISTRATIVE SKILLS OF MANAGEMENT SCIENCE

Many administrators in human-service subsystems may be found generally to be lacking in the essential management skills needed effectively to employ and administer the technologically innovative programs increasingly required in human-service areas. Such skills, in fact, may be completely lacking in the skills inventories of any of the engineering, environmental, psychological, social, medical, etc., specialists currently operating as professional human-service providers. All such traditional specialties, of course, may be required in the multidisciplinary technology; but the total complex can be orchestrated best only by one possessing the essential management skills to integrate goals, develop schedules, implement, monitor and manage mission-oriented pilot projects.[28]

Essential skill elements found in almost any well managed project today include:

Personnel Management. The ability to select appropriate personnel, and maintain a level of skill and motivation commensurate with their working consistently toward goals and subgoals of the project.

Planning. The capability of developing essential specifications and mission objectives of the human-service project through an assemblage of appropriate teams of technical specialists. Skills in scheduling time for implementation, testing and evaluation of results, cost estimating to completion through Gantt charts, etc.

Computer Use and Justification. Knowledge and ability to perform preliminary trade-off studies to establish that volume and access to invisible data in computer storage obtained through modeling and simulation will enhance planning and evaluation procedures, and result in cost saving.

Documentation. Skills in maintaining records and mission-oriented research data, and in monitoring project progress to determine problem areas and solutions.

Performance Evaluation Review Techniques and Critical Path Method (PERT/CPM). The ability to maintain close monitoring surveillance over project progress, and critical paths to completion, periodically and regularly rescheduling manpower and resources to maintain schedules to project completion.

Operations Research Studies and Review. Problem orientation skills to direct and review findings and establish costs and benefits to be derived from various alternative innovative approaches in the project, through employment of resources on a broader operational scale, based on evaluation of outcomes.

These latter appear to be among the essential skills comprising the administrative functions of a human-services project manager. Since few of these are possessed by professionals currently operating in human-service subsystems, they might serve as a sound basis for training functions design in the administrative areas. Indeed, all human-service systems could vastly benefit from such administrative skills, whether the nature of the subsystem is fragmented and lacking in goal definitions, or organized and amenable to specification of precise design parameters.

NOTES TO CHAPTER 10

1. J. Christensen and R. Mills, "What Does the Operator Do in Complex Systems?", *Human Factors Journal* 9 (August 1967) :329–40; R. Sorenson, "Manpower System Models in Personnel Allocation Research," *Human Factors Journal* 10 (April 1968) :99–106; R. Miller, "A Method for Man-Machine Task Analysis," (Wright Air Development Center, Wright-Patterson Air Force Base, Ohio, WADC 53–137, June 1953) ; C. Bennett, "The Human Factors of Work," *Human Factors Journal* 15 (June 1973) :281–88; R. Meyer et al., "Behavioral Taxonomy of Undergraduate Pilot Training Tasks and Skills: Taxonomy Refinement, Validation and Operation," *Air Force Human Resource Laboratory* TR 74–33 (III) .

2. G. Eckstrand et al., "Human Resources Engineering: A New Challenge," *Human Factors Journal* 9 (December 1967) :517–20.

3. C. Kelley, "What is Adaptive Training?", *Human Factors Journal* 11 (December 1969) :547–56; K. Alvares and C. Hulin, "Two Explanations of Temporal Changes in Ability Relationships: A Literature Review and Theoretical Analysis," *Human Factors Journal* 14 (August 1972) :295–308.

4. R. Sommer, *Personal Space. The Behavioral Basis of Design* (Englewood Cliffs, N.J.: Prentice-Hall, Inc., 1969) .

5. J. Kenney, *Police Administration*, (Springfield, Ill.: Charles C Thomas, Publisher, 1972) .

6. Space-General Corporation, "Prevention and Control of Crime and Delinquency" (El Monte, Calif.: Space-General Corp., 1965).

7. M. Pilot, "Health Manpower 1980," *Occupational Outlook Quarterly* 14 (Winter 1970) :2ff.; I. Butter, "Improved Statistics are Required—Manpower," *Hospitals* 46 (July 1972) :56–59; H. Mason, "Manpower Needs by Specialty," *Journal of the American Medical Association* 219 (March 1972) :1624ff.

8. "Periodic Reevaluation Urged of Health Manpower Needs," *Hospitals* 47 (February 1973) :109f.; J. Kralewski and P. Levine, "The Physician Manpower Question Revisited," *Hospitals* 47 (October 1973) ; "Meeting Health Manpower Needs Through More Effective Use of Allied Health Workers" (U. S. Department of Labor, Washington, D.C., 1973).

9. A. Cloner, "The Influence of System Theory in Educating Health Services Administrators. The University of Southern California Experience," *American Journal of Public Health* 60 (June 1970) :995–1005; W. Hoff, "Resolving the Health Manpower Crisis—A Systems Approach to Utilizing Personnel," *American Journal of Public Health* 61 (December 1971) :2491–99.

10. G. Albee, "Models, Myths, and Manpower," *Mental Hygience* 52 (1968) : 168–80; "Training for What?", *American Journal of Psychiatry* 128 (March 1972) : 118ff; E. Torrey, "Psychiatric Training: The SST of American Medicine" *Psychiatric Annals* 2 (February 1972) :60–71.

11. "Psychology's Manpower: The Education and Utilization of Psychologists," *American Psychologist* 27 (May 1972).

12. H. Goodwin, "Improvements Must be Managed," *Journal of Industrial Engineering*, November 1968, pp. 638–44.

13. W. Knowles et al., "Models, Measures, and Judgments in System Design," *Human Factors Journal* 11 (December 1969) :577–90.

14. J. Burgess, "Ego Involvement in the Systems Design Process," *Human Factors Journal* 12 (February 1970) :7–12; R. Taylor and E. Torrey, "The Pseudo-Regulation of American Psychiatry," *American Journal of Psychiatry* 129 (December 1972) :34–39.

15. M. Bodie and C. Sandiford, "Mental Health Associates: One Answer to the Manpower Shortage," *American Journal of Nursing* 71 (July 1971) :1395–96; A. Collins, "Natural Delivery Systems: Accessible Sources of Power for Mental Health," *American Journal of Ortho-Psychiatry* 43 (January 1973) :46–52; J. Collins, "The Paraprofessional: I. Manpower Issues in the Mental Health Field," *Hospital & Community Psychiatry* 22 (December 1971) :18–26.

16. G. Abrams and N. Greenfield, "A New Mental Health Profession," *Psychiatry* 36 (February 1973) :10–22.

17. P. Fink and H. Zerof, "Mental Health Technology: An Approach to the Manpower Problem" *American Journal of Psychiatry* 127 (February 1971) :122–25.

18. "College of Dupage to Train Human Services Generalists," *Human Services Career Center* 11 (February 1971) (201 N. Wells Street, Chicago) ; L. Sprague et al., *Teaching Health and Human Services Administration by the Case Method* (New York: Behavioral Publications, 1973).

19. W. Bower, "Recent Developments in Mental Health Manpower," *Hospital and Community Psychiatry* 21 (January 1970) :23–27; C. Brenneis and D. Laub, "Current Strains for Mental Health Trainees," *American Journal of Psychiatry* 130 (January 1973) :41–45; R. Cohen, "The Collaborative Coprofessional: Developing a New Mental Health Role," *Hospital & Community Psychiatry* 24 (April 1973) : 242–46.

20. R. Vidaver, *Developments in Human Services Education and Manpower* (New York: Behavioral Publications, 1973).

21. S. Kurland, "An Associate of Arts Program for Training Mental Health Associates," *American Journal of Public Health* 60 (June 1970):1081–90.

22. D. Nellis and J. Burgess, "An Administrative Technique for Personnel Management and Training in Mental Health," *Public Administration Review* September/October 1974, pp. 496–99.

23. Illinois Department of Mental Health, Executive Order No. 50: Mental Health Generalist Series, Springfield, Illinois, July 30, 1971.

24. The east-central region of the Illinois Department of Mental Health provides mental health services for a 16-county area with a population of approximately 760,000. It includes a resident or inpatient facility of 200 beds, with a field staff providing direct services, as well as organizing and monitoring state-funded community-owned services within the community.

25. U. S. Air Force, Military Specification: Data, Qualitative and Quantitative Personnel Requirements Information (QQPRI) MIL-D-26239A, 1961.

26. E. Fleishman, "Performance Assessment Based on an Empirically Derived Task Taxonomy," *Human Factors Journal* 9 (August 1967):349–66.

27. J. Walsh, "Technological Innovation: New Study Sponsored by NSF Takes Socioeconomic, Managerial Factors into Account," *Science* 180 (1973):846–47.

28. P. Drucker, *The Practice of Management* (New York: Harper and Row, Inc., 1954); L. Goodwin, "Middle-Class Misperceptions of the High Life Aspirations and Strong Work Ethic Held by the Welfare Poor," *American Journal of Orthopsychiatry* 43 (July 1973):554–64; J. Hage and M. Aiken, "Routine Technology, Social Structure and Organization Goals," *Administrative Science Quarterly* 4 (1969):366–76.

Part IV

Human-Service Operational Problems and Examples of Analysis

Any new approaches to the delivery of human services must of necessity be superimposed on existing services—whether these be in welfare, in the health field, fire-fighting, police work, mass transit, criminal justice, mental health, etc. Only rarely does an opportunity arise where a more total systems approach may be instrumented; this may be true even though many areas of human services could borrow directly from proven system methods of the aerospace industry. Police work, fire-fighting, mass transit, and aspects of health-care delivery may be most easily adaptable to Research and Development (R & D) methods. Other systems, viz., softer areas where man-machine definitions become more elusive, such as in child abuse prevention, etc., may be only partially adaptive to the older methods and must call upon new approaches.

Despite the lack of clear-cut definitions of operational requirements, all systems lend themselves to strict systems analytical discipline —some, of course, more readily than others. The "softer" systems, with more complex social relations, also lend themselves to the system analysis technique. More subtle problem areas, however, become evident where appropriate intervention and system design fixes must be accompanied by persuasive and manipulative strategies in the community.

175

Currently operating systems must be confronted virtually on all fronts with creative problem solving, often in the nature of "quick fixes" that will improve the operating system through careful study and implementation.

Several examples of both "hard" and "soft" systems are described. Several examples cited were those of actual operational problems for which quick-fix system solutions were sought.

11

Human-Service Operations Amenable
to Human-Factors Systems Analysis

The methods of human-factors systems analysis, as developed for man-machine or hardware systems, have been extensively described in several publications over the past two decades.[1] That such methods are adaptable to the more elusive, and perhaps complex, operations of human service systems has been widely espoused.[2] Yet, it becomes apparent that, owing to the inordinate complexity of large social systems and the dogmatic influences of vested interest groups,[3] many human-service areas do not readily lend themselves to such analyses. Unfortunately, fragmentation and disjointedness persist in these areas.

In a welfare context, a disadvantaged teen-age girl, for example, may become pregnant. She may require multiple services to overcome the lamented cycle of misfortune and dependence; prenatal medical care, nutrition guidance, guidance on being a parent, continuation of education, obstetric and pediatric care for parturition, vocational education, job placement and assistance, day care for her child, family planning, etc. Failing to receive a number of such services, she can only be expected to become a career welfare dependent.

The welfare services in practice are narrowly conceived and administered, operating in virtual isolation from each other. Indeed, such operations prompted President Nixon in the early seventies to exclaim that such human-service approaches make a mockery of our humanistic instincts, in creating disincentive to work, inequitable treatment from one geographic area to another, and a veritable mire of self-perpetuating, inappropriate, inadequate or innocuous service functions.[4]

While many human-service areas are impeded in their development, only slowly yielding to effective analytical and design methods, other such areas do indeed provide a basis for concrete analysis and design. Several areas within a number of human-service subsystems will be examined in the present chapter for potential or exemplified human-factors systems applications, and contrasted with those areas least amenable to analytical and design access. The several areas considered as lending themselves to a human-factors systems analysis have been identified in the context of surface traffic control, fire protection and health-care subsystems.

A SURFACE TRAFFIC HUMAN-FACTORS SYSTEMS ANALYSIS

During the Christmas holidays of 1973, Amtrak passenger trains from St. Louis to Chicago and from Chicago to Denver and San Francisco were delayed 13 to 24 hours. These delays were accompanied by anxiety and discomfort to the passengers who were without heat, food, and information. The transit operation was plagued with poor planning and design, and human error that ranged from inadvertently pouring water into the diesel fuel tank, running out of fuel from extended periods of idling, and the main line becoming clogged with freight train derailment.[5] These are but a few of the common faulty episodes in mass transportation resulting in public complaints that include waiting for baggage in a room without minimal seating convenience, baggage delays resulting in missing bus schedules, isolated and remote air terminals requiring inordinate travel time for access, etc.[6]

An increasing number of city management operations, because of pressures of the fuel crisis, are sometimes effectively addressing the mass transit problem as in Chicago, Dallas, Toledo, San Francisco, and Montreal and Toronto, Canada.[7] Various alternative solutions may be sought, such as staggering work hours of commuters entering and leaving the inner city to reduce congestion, to use mass transit carriers more efficiently when standing idle except at rush hours, and to make more seats available with more level morning and evening passenger volumes. Geographic distribution of population must also be considered in such planning for alternatives.[8]

Alternatives in physical modes of mass transit are also being considered, such as a personal transit automated roadway combining the speed and privacy of the automobile with the efficiency of rail transit;[9] or, employing currently operating bus transit systems more efficiently through "Dial-a-Bus" feeder services, etc.[10] Other advanced systems include high-speed tube vehicles, tracked air-cushion vehicles, magnetic levitation, fan-drive systems, etc.[11] Alternative systems are evaluated on a trade-off basis for such criteria as unique advantages and limitations, block speed, top speed, passenger capacity per hour, development status, horse power per passenger, capital costs, etc.

Increasingly such mass transit systems may come in demand, if not due to the energy crunch and depleted supplies of gasoline, then on a rational basis. In the interest of energy conservation, and reducing pollution and congestion from uncontrolled private auto usage, development of mass transit systems are likely to come under increased government sponsorship. Currently, for example, in the outlying work areas of Chicago, 70 to 80 percent of all workers drive to work, and, in 1973, over half of the workers in the city of Chicago itself drove their autos to work. In the latter case, a mass transit elevated railway and commuter trains were already available. Moreover, the majority of all private auto vehicles in such commuting tend also to be occupied by only one passenger.

It becomes evident that such massive and inefficient employment of private auto vehicles must eventually yield to the deployment of effective, efficient, and acceptable alternative modes of mass transportation. The analysis and design of such concrete alternative approaches to mass transit systems may be ideally suited to human-factors approaches. The following analytical format and rationale may serve as a basic illustration of mass transit systems and human-factors analytic approaches to design.

A SYSTEM FOR MASS TRANSIT FROM OUTLYING SUBURBS TO INNER CITY HUB AND WORK AREA

Mass transit commuting time to the inner city from a point at the extreme nominal distance of residence in an outlying suburb may not be acceptable beyond one hour. Thus, a typical passenger portal-to-portal transit mission may be described on a time line as follows:

Minutes	Function
0–10	Feeder connection from home to mass transit line
11–15	Phasing arrival for boarding main transit line inbound
16–20	Enter main transit line inbound
21–40	Transportation to inner city hub
41–50	Exit main transit line and connect with transportation to work area
51–60	Converge on work area

Subsystems may be designated as:

> Feeders to main line
> Mainline outlying connection stations

Mainline vehicle
Inner station (s)
Intown transit modes

All subsystems must, of course, be integrated to accomplish the overall mission, and, indeed, to be acceptable on a cost and efficiency basis as an alternative to the private auto.

Each of the mission phases must be thoroughly considered and subsequently be detailed for more precise man-machine functions allocation, e.g., collection of fares, and the means to accomplish these, such as using ticket booths, vending machines or other automatic collection devices. Basic criteria to be considered must also, of course, include inducing the passengers, who would otherwise drive their private autos, to utilize the system, either through natural convenience advantages, or on a mandatory basis.[12]

Thus, subsystem design criteria may include the following:

Feeder Subsystems—convenience and timeliness providing easy access and minimal congestion at outlying stations.

Main Line Outlying Connection Stations. Easy and direct access to boarding points, with minimal movement time, queuing, safety hazard and waiting time; adequate access points, vestibules for seating and standing, restrooms, easy information access and effective direction labels.

Main Line Vehicle

Entrance: minimal time to seating, convenient luggage stowage, hand holds, safe door regulation, control of inadvertent vehicle motion, etc.

Passenger Positioning: Adequate and adjustable capacity, aisle spacing, seat design comfort, adequate hand holds, standing space and restraints, etc.

Environment and Passenger Comfort: Controlled interior temperature, humidity, air currents, noise, inside-outside glare, acceleration-deceleration, vibration, riding quality and interior lighting, etc. Controlled operation in weather extremes and atmospheric effects such as salt air, etc.

Decor: Pleasing and attractive exterior, interior color and furnishings, window shapes, etc.

Emergency Procedures: Derailment prevention, fail-safe design in power failure, emergency exit provisions, etc.

Exits: Provisions for efficient, safe and rapid egress with convenient location of doors appropriately spaced for smooth flow, etc.

Operator Control Station Design: Work space and control display design optimized for efficiency and safety.

Scheduled Maintenance: Optimal and efficient maintenance routines, preventive maintenance and design for maintainability.

Inner Stations. Efficient and rapid flow of debarking and boarding passengers to main line vehicles. Minimal movement time in walking and access to street or other vehicles in network. Adequate space for seating or standing, convenient restrooms, easy information access and effective direction labels, etc.

Intown Transit Modes. Minimal queuing and waiting; rapid access and convenient, restful conveyance, etc., efficiently linked throughout the city hub with main line vehicle inner stations.

It can be seen that a full gamut of human-factors analytical procedures may be applied to such a mass transit human-service operation, which readily lends itself to such design analyses.[13]

MASS TRANSIT ANALYSIS IN CURRENT OPERATIONS—THE BARTD EXAMPLE

Human-factors systems specialists have often failed to lend their services where they may significantly contribute to public transit. This, of course, is because they are not called upon to assist, nor perhaps are they even known by public officials and administrators to possess the skills that could make a marginal system work.[14]

Mass transit presents a public service that should increasingly be found in urban and suburban areas in the planning stage or already under development. Such mass transit systems may often flounder at a marginal or submarginal operational level which appropriate human-factors/systems engineering might largely have circumvented. The human engineer's quick-fix skills, capable of alleviating many current operational problems, generally go unsolicited; nor are they aggressively extended by the human-factors engineer himself.

An analysis was made in San Francisco through state and district management contacts, by assuming the initiative in a human-factors role for a mass transit system.

The mass transit system chosen has currently proven to be excessively costly and inefficient because of what may have been only fragmented human-engineering systems analytical contributions.

Among current international and national crises, mass transit in this country is the third major concern and expenditure following defense and energy production.

In a special issue of the *National Geographic Magazine* there appeared an article entitled, "The Coming Revolution in Transportation."[15] Operational and conceptualized versions of surface traffic range from automatically controlled private automobiles to an experimental aerotrain. The latter, invented by Jean Bertin, a Frenchman, rides on an air cushion at the center of the vehicle, and is controlled in the running mode by propellers. The aerotrain prototype has a speed capacity of 240 mph.

A team, under the direction of Joseph Foa of Rensselaer Polytech, have developed a tube train with design capacity speeds of 350 mph.

The Japanese Tokaide line trains already present a practical operational vehicle. Every twenty minutes during rush hour traffic over 1,000 passengers are carried 320 miles to and from twelve commuting sites to Tokyo at speeds up to 130 mph.

Penn Central's Turbo train is also operational through Rhode Island, though operating at only half its speed capacity of 170 mph due to degraded track beds.

The article, though comprehensive and imaginative, strikes a stark note of reality, quoting from the Secretary of the Federal Department of Transportation: ". . . future innovations in transportation will have to be superimposed on a system which already exists—a system which is being expanded and being built to last for a long, long time . . ."

The reality of this observation is seen in the thousands of miles of roads for automobiles still being built across the country; while automobile manufacturers continue to stockpile automobiles—an increasing number of compacts but still the gas-burning internal combustion machines contributing both to the fuel shortage and pollution. It is seen in the thousands of miles of railroad track built in past decades though now only minimally developed and maintained resources.

It was in just such a problems context that the San Francisco mass transit development began a couple of decades ago to become the only such electric rail subway system west of the Mississippi.

The Bay Area Rapid Transit District (BARTD) was designed by a systems consulting firm; as of this date it is continuing in development. As a modern example of mass transit systems, many of its development problems may be cited in the spirit of caution for current and planned mass transit systems in other cities. Major problems include over a 60 percent overrun in costs, or underprojections not dissimilar to aerospace projects. Even in its early operational phases, the system still evidences severe operational and debugging problems merely to maintain routine operations while striving to achieve improved operational hours and range of service.

Problems that have plagued BARTD may, of course, be attributed to a great variety of circumstances ranging from strikes at the subcontractors to simply poor planning and design, not the least of which was perhaps the lack of comprehensive systems engineering.

Figure 19 presents the administrative scheme or hierarchy of BARTD's operational management. Note that this, as in a majority of operational management schemes, is only tangentially related to the adjunctive functional and problem areas. (See functional flow chart in Figure 20.)

BACKGROUND OF BARTD

BARTD began its planning as early as 1947 when an Army-Navy Joint Commission recommended a transit tube beneath the bay. The California legislature created a study commission in 1951, and, in 1957, the directors first met on an approved five-county district. An engineering consulting firm was retained in 1959 for systems design. During 1962, two of the five counties dropped out, and a three-county transit plan was adopted. A $792 million bond issue was approved by district voters for a 75-mile system. Construction began in 1964 with both district and federal money. Upon completion of the bay tube, a vehicle contract was awarded to the Rohr Corporation in July 1969, and the first production car was delivered in November of 1971.

In September of 1972, BARTD opened the first 28 miles of service. Service through the Trans-Bay San Francisco-Oakland tube began in August of 1974, with 75 miles of service completed while developments continued.

The BARTD system is perhaps the newest of mass transit operations on an operational scale in the nation. Its scope of development was based on projected populations for the seventies and the eighties. System operational goals or requirements were thus to increase the service through increasingly satisfied volumes of population. Subgoals appear to have been to automate the operation in order to reduce human-operator employment requirements. The latter, however, necessarily introduces multiple service or operational problems, as well as creating excessive computer and car maintenance demands.

The systems study of BARTD was undertaken during the summer of 1975, primarily for purposes of identifying system malfunctions

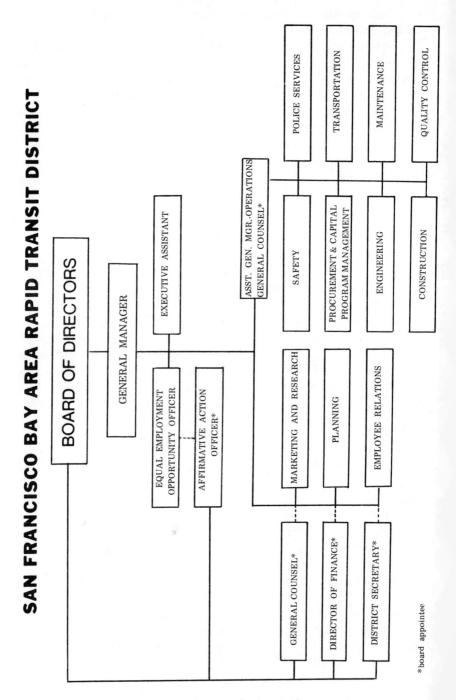

Fig. 19. BARTD Organizational Chart

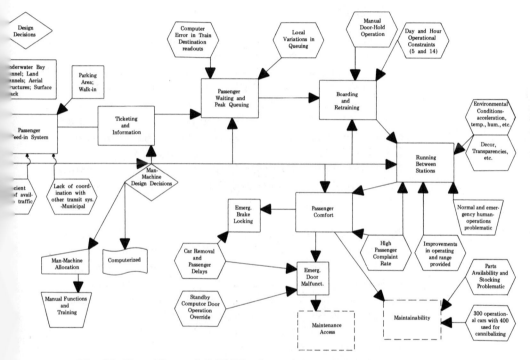

Fig. 20. Flow Chart of BARTD Operations and Problems

and man-machine problems where human-factors engineering might have paid off, or where quick-fixes might still correct errors inherent in the system. In such public service areas, the necessity for an aggressive and liberalized systems view may become essential if human-factors systems engineering is to contribute its fair share of technology to current crises. Demonstration of the feasibility of such a contribution was a secondary objective of the study.

RESULTS OF STUDY

After a broad systems study of BARTD a number of problems

were organized for discussion and design consideration. A Design Engineering Review (DEI) type of format was used, since this allows for correction of possible misconception of the problem itself, and permits a broad range of alternative solutions to be considered:

1. *Problem*: User service, while undergoing improvements and further developments, is fragmented and uncertain, and deters potential BARTD users from scheduling their use of the system.
   Recommendations: Employ all available media to inform commuters, day-time shoppers, etc., as improved and extended service becomes available.
   Reason: Such information aggressively advanced should accelerate the growth of user population.

2. *Problem*: The local and national image of BARTD is severely degraded. After detraining, comments are heard such as, ". . . every time I ride this something happens . . ." ". . . sickening ride . . ." ". . . fiasco . . ." etc. The passengers are not informed about trouble and delays, thus often becoming hostile and angry with BARTD.
   Recommendation: Drivers should be trained to explain categorical troubles and expected delay times to the passengers, or selected tapes might be actuated for play over the public address system.
   Reason: Knowledge of reasons for trouble and delays might help the passengers to identify with the situation, and, with the advent of improved operation, have a better feeling about BARTD. BARTD's image might also be improved if effort were made to explain its growing pains and the projected service in its complete configuration.

3. *Problem*: The glass at the door of the driver's compartment attracts passengers to view the controls and gather around the door. The driver's ingress and egress is blocked during emergency periods, further retarding the operation.
   Recommendation: Replace the glass in the door with a one-way glass screen, i.e., a mirror on the passenger side and transparency on the driver's side.
   Reason: Curious passengers will then tend not to congregate about the driver's compartment, making for easier access to the compartment by the driver, etc.

4. *Problem*: The technology of mass transit in the United States appears to be limited compared to that of countries in Europe.
   Recommendation: Make conditions for federal financial assis-

tance contingent upon the free and cooperative exchange of state-of-the-art information both inside and, as available, outside the country.

Reason: Mass transit technology may best be advanced by high interaction rates among engineers and scientists.

5. *Problem*: The fare collection system provides maximum change for single dollars only, with five-dollar bills frequently lost by the clients, further contributing to their dissatisfaction with BARTD.

Recommendation: Post caution signs to warn customers of the loss of five-dollar bills when inserted, or improve the "reject" function of the change machine.

Reason: Newcomers are the ones most likely to lose their five-dollar bills. When necessary to read instructions, a caution note, given prominent display, is likely to be heeded.

6. *Problem*: The All-Transit system—BARTD, buses, trains, trolleys, etc., operate independently, often duplicating traveled routes, while sometimes not covering some areas of the district.

Recommendation: Develop master coverage plan as a comprehensive transit total system, arranging schedules, routes, etc., to avoid duplication, while covering all essential areas.

Reason: Economy, efficiency, and improved service should result.

7. *Problem*: Mass transit systems often run excessively over in costs, to the chagrin and disgust of taxpayers.

Recommendation: The need to take a more comprehensive systems management approach is indicated, where planning otherwise falls short in such contingencies as labor problems and strikes, unforeseen technical problems, etc.

Reason: Comprehensive and realistic planning should avoid much resentment and reduce the severity of the public relations problem.

8. *Problem*: Numerous complaints can be expected when the car components mechanically become loose with wear and start noisy vibrations.

Recommendation: Design baffle systems to dampen noise when the buildup is expected to occur.

Reason: Prevention of excessive noise will further enhance consumer acceptance and improve public relations.

9. *Problem*: Lack of necessary spare parts often holds up repair work.

Recommendation: Complete time-to-failure studies as a basis for maintaining spare parts stocking levels.

Reason: Since lack of necessary spare parts in maintenance is due in part to the new state-of-the-art and lack of operational experience on failure rates, compiling time-to-failure data, as a supplier data requirement, could later prevent excessive down time.

10. *Problem*: The computer may erroneously reverse readout of the terminal point of the train. This confuses passengers and may further result in dissatisfaction with BARTD.

    Recommendation: Introduce "fail-blank" circuits when such failures occur, leaving the readout blank. Provide backup radio transmission for the drivers over the local Public Address system at the station to announce the final destination of the train. This might serve as both a backup and a means to achieve reliability by redundancy.

    Reason: A significant annoyance may be further removed from passenger service in quick-fixing a confusing malfunction which further causes passenger discontent with BARTD.

11. *Problem*: Passengers queue up at doorways to BARTD cars from previous bus boarding patterns, thus blocking passage for those who need to enter and exit at farther points along a ten-car line.

    Recommendation: Post signs such as, "Please use all doors to enter train," etc.

    Reason: Signs may, with some reminding assistance from station personnel, break down the old boarding patterns.

12. *Problem*: Spray paint is sometimes used for making graffiti in the underground stations, and at surface areas vandalism is evidenced by rock throwing, cracking of expensive windows, etc.

    Recommendations: As a public relations function attempt to instill pride in the system at all levels, through the schools, newspapers, etc., and through generally improved services.

    Reason: Vandalism appears to be a universal problem in all such train operations. Through strategic public relations action it might be possible to alleviate some varieties.

13. *Problem*: When the system was ready to operate, there was literally no one who had complete knowledge of the complex total system operation of BARTD. There are few likely currently to have such knowledge.

    Recommendation: In future such systems, a systems group should be formed, responsible for imparting knowledge of the complete system to all operational personnel.

Reason: Complete systems knowledge by all key personnel may be essential for smooth performance.

14. *Problem*: Car components such as doors and brakes have in the past had high failure rates. Through improved reliability, failure frequency has decreased, though failures still occur and result in breakdowns. Such failures harass the passengers with excessive delays, anxiety, uncertainties about their safety, etc.

    Recommendation: In the future, such systems should be designed to provide for such high failure rates through an informed passenger public, expedient computer override emergency procedures, and time line studies for minimal delays.

    Reason: Efficient passenger service (and of course passenger safety) should be of primary management concern. Such system studies of failures, and of emergency procedures, may greatly enhance passenger satisfaction with BARTD.

15. *Problem*: Human operator tasks have been only nominally introduced apparently without proper man-machine functions allocation procedures. The human operator now: (a) observes track clearance, (b) monitors annunciator panel, (c) observes platform for passenger clearance, (d) holds doors open until passengers are observably clear, (e) radios emergencies to central stations. The operator is required to take only minimal action in emergency situations, such as brake or door malfunctions.

    Recommendations: Study each task for more optimal allocation of human task functions: (a) brake malfunction—now if one car has a brake failure, the entire train must be evacuated and taken to nearest station for a replacement car. Extensive passenger delays are incurred. The driver could make judgments, with proper training, as to the extent of free wheeling safely permissible, and given operational aids to make such judgments, with time-consuming repairs to be accomplished at less demanding periods for passenger service; (b) door malfunction now requires much the same procedure as brake failure. A standby operator may be called in, and, with manual override provisions incorporated, could close doors manually until a more propitious time for repairs becomes available.

    Reason: Improved passenger service may become possible by optimizing for malfunction and manual override action.

16. *Problem*: Maintainability may often be a problem in accessibility of certain parts, as in the door mechanism in which skin over door must be removed for access to failed components in an

excessive labor- and time-consuming operation.

Recommendation: Study maintainability tasks more extensively, with an eye for improved access, minimal down time, etc. In the case of the doors, retro-fitting of door structures may prove to be a time- and money-saving operation when the total trade-offs are considered.

Reason: Improved maintainability, such as in door maintainability, may reduce overall costs of maintenance, and reduce down time of single vehicle. This could become increasingly important as service and car usage is extended.

17. *Problem*: The frequent total system breakdown may require an extensive reliability analysis for parallel or series component design.

Recommendation: Introduce an increasing number of parallel components to improve reliability and, as much as possible, eliminate the series components.

Reason: Reliability of components in series is a simple multiplicative function. For example, if all components have a probability of functioning properly by as much as 99 percent of the time, the reliability (R) is as follows:

$$R = C_1 \times C_2 \ldots C_N$$

With 100 such components in series, the reliability is reduced to one in ten failures. Four hundred components in series would reduce reliability to almost certain failure. By contrast, parallel components are essentially backup functions, e.g., operator manual override of an automatic braking system; thus,

$$R = [1-(1-r)^m]^n$$

where m = number of components in parallel,

n = number of functions, and

r = the reliability of individual components established independently.

Components in parallel may have low individual reliability, but as backup functions they enhance overall reliability. Thus, if two parallel components each have an individual reliability of .90, then $R = [1-(1-.90)^2]^1 = [1-(.01)]^1 = .99$, or only one failure in 100 runs.

QUICK-FIXING MASS TRANSIT SHORTCOMINGS

The design and development of mass transit systems in cities throughout the United States may be expected to become an ever increasing systems problem in light of fuel crises and decline of the individual automobile.

Failure to employ a total systems approach could result in frequent breakdown of service, inefficient operation and prolonged delays for the passengers, severely complicating the public relations problems. The BARTD system evidences a number of such problems. A brief analysis of the BARTD system indicated that early human-factors systems design considerations may have averted many of the current problems. Implementation of a number of quick-fix procedures in current and future mass transit systems, on the other hand, could also possibly alleviate many of the operational difficulties. Initiative and erudition on the part of the human-factors systems engineer appear to be most needed if their technology is finally to be effectively introduced to a cost-conscious public service management.

A FIRE-PROTECTION HUMAN-SERVICE SUBSYSTEM

A priori analysis of a fire protection system also indicates an amenable, though uncommitted, potential for systems and human-factors analytical approaches. A typical mission analysis, for example, assuming a maximum 15-minute reaction time requirement, might be outlined as follows:

Minutes	Function
0	Fire alert
0–1	Signal alert to station
2–3	Scramble
4–11	Locate and move to area of fire
12–13	Determine scope of fire and persons in danger
14–15	Deploy rescue and fire-extinguishing equipment

Subsequent phases would involve rescue and extinguishing, securing area, reassembling, and return to station. Such a human-service system may be readily seen as lending itself to human-factors analyses, with such subsystem components as detection, signal communication, station rescue, fire control and vehicular components, an information control subsystem for locating the scene of the fire, and a rapid-assessment subsystem for efficient deployment of the rescue and fire-extinguishing equipment at the scene of the fire. The potential use of such systems and human-factors analyses should hold major promise for greatly enhancing the efficiency and effectiveness of the fire-control subsystem.

A HEALTH TRAUMA FAST-REACTION SERVICE SYSTEM

Among health service operations that lend themselves to a human-factors system analysis are those emergency services designated as "Trauma Centers." Such centers have been established for several regions throughout the country, and, most notably, throughout the State of Illinois. The essential service specification is a rapid reaction to, or on, the site of injury, e.g., highway or farm accidents, building or resident fires, etc., or physiologic accidents, e.g., heart attacks, diabetic seizures, suicide attempts, etc. System design requirements may be generally described as providing the most qualified care in the most expeditious way possible.

A representative mission of the Trauma Center may be described as follows:

Time, zero plus 10 seconds
A highway accident occurs between truck and automobile, with auto rollover and severe damage. Auto occupant injury indeterminate; two passengers, both unconscious.

Plus 1 minute, 30 seconds (or, if random, indeterminate)
Truck driver surveys damage and radios highway patrol or police (or the latter are otherwise notified).

Plus 5 minutes
Police arrive on scene.

Plus 5 minutes, 30 seconds
Injured occupants are carefully removed. Stabilization routines are initated.

Plus 15 to 30 minutes
Injured occupants are transported to emergency treatment center.

Plus 30 to 115 minutes
Care and treatment are provided at emergency treatment center.

Plus 115 minutes to four hours
One patient is transported to advanced treatment center as needed.

Plus four hours to x number of days
Advanced treatment is provided as necessary.

Various functions may be time shared to enhance operational efficiency, such as emergency medical technicians monitoring police calls to arrive on the scene shortly after police arrive, communication with treatment center for appropriate standby equipment and personnel during transport phase, etc.

Three subsystems of the Trauma Service System may be designated:

1. Stabilization

2. Emergency treatment
3. Treatment network

The stabilization subsystem consists of transportation, manpower, skills and training, and communication components directed toward the accomplishment of sustaining functions. Stabilization essentially concerns the prevention of further injury and/or providing immediate attention to prevent more permanent damage or death, e.g., setting fractures, arresting bleeding, defibrillation of heart seizures, etc. This phase includes the imparting of skills to specialized technicians and such primary response personnel as police, firemen, families of heart attack victims, etc. Human-factors analysis may contribute in the area of task and training functions analysis, design of emergency measurement and control instruments, and communication equipment to relate with more expert medical personnel at the emergency treatment center.

The emergency treatment subsystem requires equipment, manpower, and skills sufficient to promote short-term healing and correction. The victims are rapidly moved from the site of injury after stabilization to be met with standby equipment and personnel for preliminary treatment. Centers may be designated "comprehensive," "basic," or "standby," according to level-of-treatment facilities and equipment available. Long-range rapid transport equipment, such as heliports, are also available to facilitate transfer to those facilities with more advanced equipment or specialization, e.g., burn centers.

A treatment network subsystem consists of communication equipment and functional service provisions to accomplish advanced treatment required for more permanent and refined correction and healing of the victims.

In this type of health service specification, human-factors analyses are thus nicely suited to the development of design criteria, coordination and integration of equipment and personnel interfaces to accomplish total system goals.

DESIGN OF A RECREATIONAL SYSTEM

A complete systems design approach to recreational systems or subsystems may also be readily applicable. An Australian minister, engaged in an internship in pastoral counseling, presented the system problem of providing a recreational program for the elderly in his congregation. Though this systems design problem does not lend itself to a nice time-line analysis, it is amenable to the system analytical procedures as outlined in chapter 3.

Introducing a systems design approach to this pastoral counselor brought to bear his substantive knowledge of problems of the elderly in his congregation in a fashion that was to integrate resources most effectively for total system planning.

The design goal was formulated in fairly straightforward terms—to design and develop a recreational program for the elderly in the church congregation to accomplish 50 percent total participation during the first six months. Subsequent goals were to be developed or reshaped in the feedback process on design elements. For example, evening recreational pursuits for the elderly may conflict with teen-ager programs—noise from the basketball court, teen-agers laughing and chasing about the halls, etc., may have interfered with the elderly recreational activities, thus discouraging attendance. While this may not have been evident until several months into the program, new design goals, or delays in initial goal accomplishment, may have been indicated.[16]

With clearly formulated goals, the following design procedures were instituted (refer to chapter 3):

1. *Assembling Design Expertise*—One systems-oriented specialist was sufficient working with the pastoral counselor in addressing a system problem within his own congregation. Employing the currently available church facilities and resources was generally sufficient, while the recreational system design was essentially under a single jurisdictional decision maker, viz., the pastoral counselor and his ministry.

2. *System Requirements Review*—Review in addressing recreation for the elderly consisted simply of cataloging the various possible programs and activities available. These included, in addition to general socializing, "Skilled," "Intellectual," and "Chance" games.[17]

3. *Feasibility Planning*—This phase of analysis consists of evaluating the various possible alternatives and preparing a mission plan. In the latter case this is simply projecting a series of activities making up a complete recreational session.

Alternatives: Major constraints lie in resources available, and in the degree of vigorous participation that might be expected of this age group. Feasible activities for the elderly might be generally grouped as follows:

Skilled—shuffleboard, billiards, dancing, etc.

Intellectual—charades, scrabble and other word games, etc.

Chance—various card games, monopoly, bingo, etc.

Trade-offs in the selection of several of these alternatives, of course, hinged about cost constraints. Some of the necessary equipment, such as scrabble, cards, monopoly sets, etc., were currently available in church stock. Others might require purchase out of limited resources. Volunteers, or professional services, as in dancing instruction or square dance calling, etc., would also present constraints. Majors contraints may also be indicated in validating the type of activities that would attract the greatest number of this population.

Mission Analysis: Projected sequences may be described as follows:
- —Program communication to elderly
   announcements
   bulletins
   personal friends
   phone calls
- —Transportation at appointed hour
- —Leadership development and selection
- —Games and activities
- —Provisions for space accommodations of 40 elderly persons in all activities (games, food preparation, serving, dining, socializing etc.)
- —Maintaining and creating further interest in attendance
- —Conducting day sessions
- —Conducting evening sessions
- —Providing transportation home
- —Providing agency referral as needed

Preliminary Testing or Modeling: Simplex models may often be of value in planning and evaluating recreational sessions. Flow charts, for example, may serve to provide overviews of proposed systems as well as those in operation. Problem areas may be identified and described at each step or mission segment.

4. *Determine Subgroups or Subsystems*—Subsystems or groups of elements necessary in developing or sustaining a recreational system operation. These would include the following subsystems:
- —Transportation
- —Communication
- —Space facilities
- —Gaming or recreational components
- —Cooking or provisions for refreshments
- —Dining facilities
- —Socializing facilities
- —Record keeping

Functions Analysis: Subsystems and mission requirements must be translated into definitive functions, such as carrying or moving a group of up to 40 elderly people over a distance of three to four miles from home to church, afternoon or evenings. Such a breakdown may include only a level about which specific determinations of methods may be made, such as church buses, private volunteer autos church members might drive and pick up, etc. *Functions allocation*: In the total systems process, functions may be identified which could be best handled by a machine function, or which may be compared on a cost-benefit basis with manual processing. For example, record keeping of attendance, special problems, next of kin, etc., may be assigned as a machine function, i.e., computerized, card punch and sort system, etc. In rational analysis, however, a population volume of 40 elderly would appear a priori to be most economically and beneficially handled manually.

5. *Population Served*—The population was described as an elderly group of approximately 40 individuals between 65 and 80 years of age. Here several variables must be considered, such as general physical condition, skills available in the group, degree of mobility and athletic inclination, preferences through group surveys, etc.[18]

6. *Assembling Pertinent Human-Factors Data*—Special requirements in environment, processing, information, the flow of information (form, content, elimination of redundancy, etc.), special aids required for the enfeebled, etc., recreational areas and furnishings, provisions for easy cleanup and maintenance after each session, e.g., group self-serving and cleanup, etc.

7. *Time-line Analysis*—Though such a recreational system does not clearly lend itself to a second-by-second or even minute-by-minute analysis, some form of time analysis may be helpful to determine conditions for full participation, e.g., duration of attention span and interest by activity, etc.

8. *Required Skills*—Apart from normally required and available operational skills, such as secretarial services, etc., the recreational skills of the elderly may be the major focus. An inventory of such skills, such as in dancing, singing, musical instrument proficiency, etc., may be necessary for improved participation. Certain members of the group may be selected for leadership tasks, if recreational activities indicate such requirements; or church members may be solicited for volunteer work; where necessary, leadership training, recreational skills in dancing, etc., may be necessary.

9. *Evaluation*—The feedback loop, or ongoing monitoring of participation data, must be imparted to the pastoral counselor in

design/redesign of the program for optimal participation. Theoretical evaluation may be based on a rationale to enhance the life quality of the elderly.[19] Criteria such as these may be derived from surveys or possibly improved longevity, etc. Monthly program review on the basis of participation measures, expenditures and attendance, etc., may serve as pertinent management information. Likewise, long-term review may also provide a basis for program improvement in accomplishing attendance and other goals.

Recreation and other social systems may thus (to a greater or lesser extent depending largely upon jurisdictional control) lend themselves to total systems development.

EVASIVE "NON-SYSTEM" HUMAN SERVICE SUBSYSTEMS

While various subsystem components may lend themselves to systems and human-factors analyses, such as the Trauma Center of a health-care subsystem previously described, the majority aspects of several human-service subsystems may be simply so fragmented and resistant to change that they may be essentially described as having a "nonsystem" character. Those seemingly most readily lending themselves to system and human-factors analysis descriptions would seem to be predominant aspects of industry and commerce, surface traffic, air traffic, fire protection, and sanitation subsystems. Those most resistant, chiefly for reasons of more formidable professional vested interest and political boundaries, would seem to be the domiciliary, law enforcement and criminal justice, health, mental health, and public welfare subsystems.

Health care, for example, is currently dominated in policy and practice by private practicing physicians. Health Maintenance Organizations (HMOs), as systems of health care involving (1) an insurance or collective funding source, (2) a participating body of consumers, and (3) cooperating and sharing physicians, are extremely slow to develop. This is perhaps due to the physicians' reluctance to yield their time, effort, and resources at the expense of private practice and compromised earning power. Current concern with progress in health-care delivery is often predicated on multidisciplinary approaches.[20] These, however, are definitively impeded by the established but artificial boundaries among health-care professionals. The persistent failure to develop an effective planning and decision-making authority thus continues to fragment the health-care subsystem.[21]

Mental health, the subsystem in which this writer has participated for the better part of a decade, is no less evasive and fragmented. In fact, the force of vested credentialism in this subsystem is of such demanding proportions as to focus on fragmented approaches while often compromising effectiveness and splintering necessary services.[22]

The public welfare subsystem has been one about which much local and national public concern is developing. Major national expenditures and the issue of welfare rights continue to draw attention to irrelevant and fragmented features of the subsystem. The governor of Illinois, during fiscal year 1974, for example, created several major issues revolving about the eligibility and deserving-character of welfare recipients. The usual excuse or rationale for "cheaters" was presented by agency heads that if more caseworkers were available, cheating could be reduced. Only infrequently are the more central system issues brought into focus such as the system's inflexibility in solving recipient's specific problems, or the lack of direction in state public-aid programs. While some few observers see the only solution as a massive overhauling of the entire program, the great majority continue to address fragmented and ineffective non-system solutions.

Human service subsystems, even those with obvious and agreed-upon parametric solutions, are by-and-large subject to only partial analytic solutions. Systems and human factors expertise, however, may continue to promote the open-systems persuasion, while assisting in interim and politic solutions that continue of necessity to be fragmented.

NOTES TO CHAPTER 11

1. R. Miller, "Some Working Concepts of Systems Analysis," *American Institute for Research* (Pittsburgh, Pa., 1954); A. Shapero and C. Bates, *A Method for Performing Human Engineering Analysis of Weapons Systems,* Wright Air Development Center, WADC Technical Report No. 59–784, September 1959; D. McRuer and E. Krendel, "The Man-Machine System Concept," *Proceedings, IRE* 50 (May 1962):1117–23; K. DeGreene, *Systems Psychology* (New York: McGraw-Hill Book Company, Inc., 1970); E. McCormick, *Human Factors Engineering* (New York: McGraw-Hill Book Company, Inc., 1970); K. DeGreene, *Sociotechnical Systems: Factors in Analysis, Design, and Management* (Englewood Cliffs, N.J.: Prentice-Hall, Inc., 1973).

2. "Aerospace Technology and Urban Systems," *Industrial Research,* April 1968, pp. 85–90; B. Schriever, "Systems Management and the Cities" (Paper presented at the Division of Engineering, New York Academy of Sciences, May 8, 1968); H. Millendorfer and E. Attinger, "Global System Dynamics," *Medical Care* 6 (December 1968):467–89; H. Holder and R. Dixon, "Delivery of Mental Health Services in the City of the Future," *American Behavioral Scientist* 14 (July 1971):893–908.

3. F. Baker, "Community Mental Health Ideology, Dogmatism, and Political-Economic Conservatism," *Community Mental Health Journal* 5, no. 6 (1969) :433–36; H. Holder and R. Dixon, "Delivery of Mental Health Services in the City of the Future."

4. J. Burgess, "A Goals Versus Process Orientation in Community Mental Health," *Social Psychiatry* 10 (1975) :9–13; F. Carlucci, "The Future Outlook for Delivery of Human Services," *Health Services Reports* 88 (December 1973) :891–93.

5. "Trains Stall in State," Associated Press, January 2, 1974.

6. W. Owens, *The Metropolitan Transportation Problem* (Washington, D.C.: The Brookings Institution, 1956) ; idem, *The Accessible City* (Washington, D.C.: The Brookings Institution, 1971) .

7. Ontario Department of Transportation and Communication, "A Study for the Selection of an Intermediate Capacity Public Transit System" (Ontario, Canada, 1972) ; P. Barnes, "Is BART Any Way to Run a Railroad? So-So Rapid Transit," *The New Republic*, September 1, 1973, pp. 18ff.; "Transportation," *Science News* 104 (1973) :70f.

8. F. Rossini and T. Tanner, "Transportation, Communication and Population Distribution," *Proceedings of the Sixteenth Annual Meeting of the Human Factors Society*, Human Factors Society, P.O. Box 1369, Santa Monica, Calif., 1972.

9. W. Hamilton and D. Nance, "Systems Analysis of Urban Transportation," *Scientific American* 221 (July 1969) :19–27.

10. Ontario Department of Transportation and Communication, "A Study for the Selection. . . ."

11. L. Hoag and S. Adams, "Human Factors in Urban Transportation Systems," *Proceedings of the Seventeenth Annual Meeting of the Human Factors Society*, Human Factors Society, P. O. Box 1369, Santa Monica, Calif., 1973.

12. C. Sundberg and M. Ferar, "Design of Rapid Transit Equipment for the San Francisco Bay Area Rapid Transit System," *Human Factors Journal* 8 (August 1966) :339–46.

13. Transportation Research Institute, "High Speed Ground Transportation" (Pittsburgh, Pa.: Carnegie-Mellon University, 1969) ; Hoag and Adams, "Human Factors in Urban Transportation Systems"; F. Oberman, "An Approach for Reflective Consumer/User Information into Systems Design," *Proceedings of the Seventeenth Annual Meeting of the Human Factors Society*, Human Factors Society, Box 1369, Santa Monica, Calif., 1973.

14. P. Kyropoulos, "Human Factors for the Masses," *Human Factors Society Bulletin* 14 (1971) :3–4; R. McFarland, "Society President Discusses Human Factors During the Next Decade," *Human Factors Society Bulletin* 13 (1970) :1–3; "Trouble in Mass Transit. Why Can't the People who put a Man on the Moon get You Downtown?", *Consumer Reports*, March 1975, pp. 190–95.

15. F. Appel, "The Coming Revolution in Transportation," *National Geographic* 136 (1969) :301–41.

16. The general systems management procedure follows the iterative sequence outlined in the diagram of Figure 6.

17. J. Levy, "An Applied Intersystem Congruence Model of Play, Recreation, and Leisure," *Human Factors Journal* 15 (1974) :545–57.

18. L. Reissman, "Class, Leisure and Social Participation," *American Sociological Review* 19 (1954) :76–84; F. Robbins, *The Sociology of Play, Recreation and Leisure Time* (Dubuque, Iowa: W. C. Brown Company, Publishers, 1955) ; B. Sutton-Smith et al., "Game Involvement in Adults," *The Journal of Social Psychology*

29 (1956) :367–77.

19. E. Staley and N. Miller, "Leisure and the Quality of Life . . . a New Ethic for the 70's and Beyond" (The American Association for Health, Physical Education and Recreation, 1972) .

20. H. Adelman, "System Analysis and Planning for Public Health Care in the City of New York," *Archives of Environmental Health* 16 (February 1968) :258–63; A. Bennett, "Systems Engineering," *Hospitals* 43 (April 1, 1969) :171–74; H. Buzzell, "A New National Strategy to Make Health Services Flexible and Responsive," *Health Services Reports* 88 (December 1973) :894–97.

21. C. Flagle, "The Role of Simulation in the Health Services," *American Journal of Public Health* 60 (December 1970) :2386–94; E. Gardner and J. Snipe, "Toward the Coordination and Integration of Personal Health Services," *American Journal of Public Health* 60 (November 1970) :2068–78.

22. A. Graziano, "In the Mental Health Industry, Illness Is Our Most Important Product," *Psychology Today*, January 1972, pp. 3ff.; W. Fisher et al., *Power, Greed and Stupidity in the Mental Health Racket* (Philadelphia: The Westminister Press, 1973) ; J. Burgess, "Who Has the Administrative Skills in Mental Health?", *Public Administration Review*, March/April 1974, pp. 164–67.

12

Human-Factors Analysis of a Rehabilitation System for Chronic Mental Patients

A systems engineering project was also recently undertaken on a community lodge system for mental patients. A description is included here since it presents a comprehensive illustration of total analysis for design of an open system[1] relating to the community. The gist of the project was to arrest the vicious recycling or extended lengths of mental-hospital stay by "chronic" mental patients. This was basically to be accomplished by developing socially cohesive cohort groups within the hospital, improving their general living and vocational skills, and setting up an ongoing lodge for continuing cohort participation in the open community. Design and operation was thus oriented to definitive outcome criteria that permitted a detailed system analytical approach.

METHODS OF ANALYSIS

Standard adaptable methods of human-factors systems engineering were employed in the analysis to develop design criteria and implementation procedures. *System Operational Requirements* were determined, a priori, on a rational basis and from consensus of operational staff. *Subsystems* were then identified as basic operational components of the system with operational conditions identified. A *mission's analysis* was then completed to identify and isolate specific operational problems for analysis and criteria development. Required mission operations were then specified for *documented* studies of rationale and state-of-the-art *design criteria development. Design and development* schedules were prepared and *budget forecasts* completed. A plan for an evaluative feedback operation was also completed for system improvement during the course of ongoing operations.

System Operational Requirements (SOR)

The general requirement is specified as reducing the recurrence of state facilities admissions by heavy users through overall living skills improvement and mutual supportive operations of the patients themselves. Mutual support is to be accomplished through social cohesiveness developed in an initial inpatient contact situation, and, through continuing interaction and the sharpening of subsidiary job, social and living skills, then to transfer and maintain the entire supportive group in a community situation. General measurable criteria for evaluating system operational effectiveness shall include reduced readmissions to inpatient facilities, increased lengths of community tenure, improved performance in productive work, increased interactive social participation and enhanced overall life quality.

Subsystem Components

In order to accomplish the SOR as described above, four essential functional aggregates are indicated to be designated as subsystems: (1) a cohort-gathering subsystem, (2) a living-skills acquisition subsystem, (3) a vocational-skills acquisition subsystem and (4) a community supportive-group subsystem. Figure 21 presents a block diagram of the subsystem components.

The cohort-gathering component physically employs the Adult Resident Unit at the Meyer Center, or community mental health centers and general hospital psychiatric wards. The gathering function explicitly applies standard criteria for "heavy user" selection in cohorting.

The living-skills acquisition subsystem physically involves the operation of the Meyer Rehabilitation Unit, while the vocational skills-acquisition Subsystem is the microcosm of a factory physically managed within the Meyer facility. Community commerce and industry become involved, as an extra system component, in factory design and marketing, and the Division of Vocational Rehabilitation adjunctively operates in training and job placement.

The community supportive subsystem consists of a nonresidential lodge organized out of the Meyer Rehabilitation Unit. The patients are provided financial assistance as required and situated in group homes or private living facilities, to support each other mutually in living and work situations through cohesive and extensive participation in the lodge. The community mental health clinic is adjunctively employed for patient support in the community.

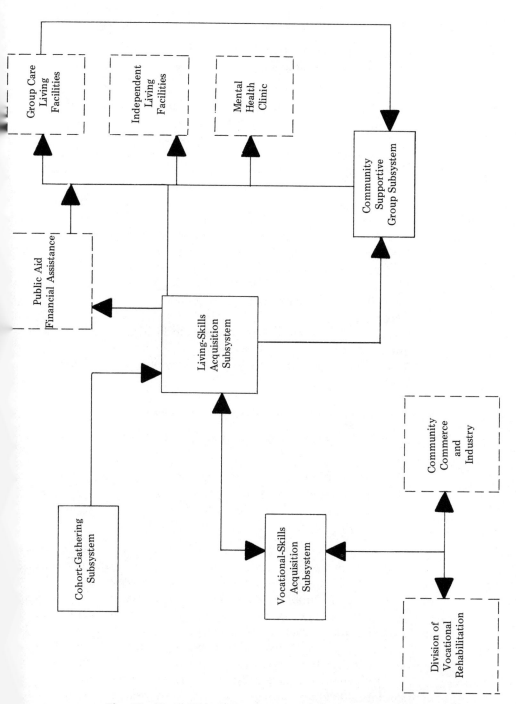

Fig. 21. Heavy-User Rehabilitation Subsystem Components

Goals for the total system include social-occupational and subsystem components as follows:

1. Hundred percent rate of lodge participation by cohorts each week.
2. Social cohesiveness among lodge members with measurable
   a. group decision-making effectiveness
   b. group problem-solving effectiveness
   c. mutual tolerance, support and advocacy among members, socially and vocationally
   d. improved "self-image" among members
   e. leadership effectiveness among members holding lodge meetings with parliamentary format
   f. membership participation in social activity (games, sports, hobbies) day and evening
   g. membership autonomy of operation with minimal staff participation
3. Contributions to community services by lodge members as a group.
4. Reduced rates of patient return to state facilities.
5. Improved work performance in quality and duration.
6. Within one year transferring lodge sponsorship to community agency responsibility.

Mission Analysis and Design Criteria Development

A mission of the rehabilitation system may be outlined in four phases as follows:

A. Gathering Heavy Users, and Preholding for Cohorts
B. Indoctrination in Living Skills
C. Indoctrination in Vocational Skills
D. Maintaining Mutual Support in the Community

Phase A. *Gathering and Preholding*—This phase entails the intake, screening and preholding functions for the heavy institutional-user population. Criteria for heavy users were first established as more than two prior institutional admissions and/or longer than 2½ years stay in any institution. Screening criteria also included family and social ties for site of community participation.

Criteria for the gathering of cohorts were then established as social-psychological functions.

Size of Cohort: In general, cohort groups of five were selected as optimal for individual participation, to prevent isolation when disagreement occurs while the dissension of one member does not jeopardize group security, and at the same time a high degree of leader influence is possible. Individual satisfaction is also more likely in work and discussion groups of five.[2] Groups of *10* appear to be optimal for supervision, as in a factory operation, i.e., there should be at least *two* supervisors for a factory operation of *20*. Larger groups, e.g., over *36*, may be acceptable for a lodge situation, where sub-grouping will occur or be sustained, while greater tolerance is indicated for leadership.

Cohort groupings of five were thus established as optimal for gathering functions, with the factory in groups of *10* and the community lodges running between 30 and 40 members.

The rate at which "heavy-user" groups of five accumulated in the catchment area was approximately 2.5 per week, or one cohort every 14 days. Figure 22 illustrates the operational flow of cohorts of five through the four phases of the rehabilitation system.

Other functional problems in the gathering phase included breakdown in the application of "heavy-user" criteria by preholding staff, i.e., staff tended to personalize and refuse to select qualified patients for extraneous reasons. This required the enforcing of standards and/or allocation of screening to a machine-processing function.

Phase B. *Indoctrinating Living Skills.* Functions in this mission phase include social cohesion development of cohorts, individual living skills assessment, individual living skills development, and community placement for lodge participation and community support, all directed toward the accomplishment of the SOR outcome goals.

Social Cohesion Development. Social cohesion must be developed and maintained for operation of mutual community support by cohort members after discharge. Cohesive functions in mutual support by cohort members are designed: (1) to create a reference group for problem solving; (2) to achieve mutual responsibility to each other, for self and for others; (3) achieve satisfaction of members through group autonomy. Individuals need also to participate in meaningful decision-making activities in deciding their own fate.[3]

Design study data also indicate that those who remain out of the hospital tend to be those with socially supportive living situations. A successful hospital social system must be moved into the community where the task groups are presented with problems of maintaining themselves in a productive and supportive community situation. Readmission rates should be reduced, employment increased, and life situations enhanced.[4]

Cohort cohesiveness must operate in such a way for the individual patient: (1) to learn to get along with other people, (2) to develop

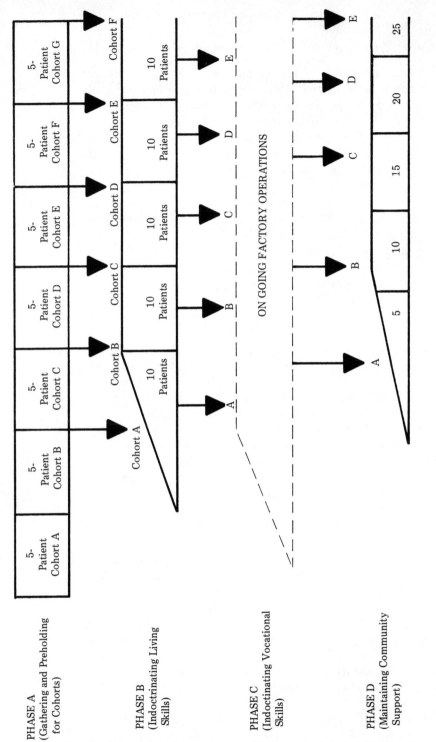

Fig. 22. Cohort Movement in a Social-Vocational Rehabilitation System

and encourage his social skills, interests and hobbies in overcoming social helplessness in the simplest contacts with others, (3) to develop his self-confidence in improving the picture of his self-image, (4) to move into appropriate social activities of the community at large, such as church fellowships or outside friends. The membership should cultivate common interests, as much as possible, and each member should learn to tolerate the eccentric behaviors of the other members.[5]

Development will thus require cohort organization and progessive cohesiveness in various functional situations as follows:

—group living as inpatient residents
—meal time in cafeteria
—group meetings as inpatient residents
—education and training classes
—recreational and social activities as inpatient residents
—vocational training and factory work
—post-discharge nonresidential lodge

Organizing and developing cohort cohesiveness basically requires bringing the patients together and keeping them together in these various situations. Staff functions may be differentially deployed to accomplish this, or rules and routines may be developed to minimize staff participation. At meals, for example, patients eat in the common cafeteria with tables pulled together. Cohort members may sit by themselves, while a large staff effort may be required to induce them to sit at a common table. An alternative may be to provide a reserved and restricted table with established rules that this is the only table available to them to be enforced by cafeteria personnel. Thus, a lesser staff level is required to maintain togetherness in the meal situation.

Other situations in group meetings, etc., may employ state-of-the-art design criteria for cohesiveness development in the context of small group theory in decision making and problem solving, mutual tolerance and support by cohort members, and functional group autonomy of large and small groups.

Cohesiveness Through Group Decision-Making. The minimal requirements for cohesiveness in cohort decision-making is to accomplish agreement. Agreement does not, of course, mean that each cohort member agrees for the same reason. One cohort member, for example, may agree to an activity because he wishes to avoid ward duties, while another member may agree to the same activity because he can be in town to do some shopping, etc.[6] Conditions for consensus and cohesiveness through decision-making include:

—a positive affective atmosphere

—intelligibility of topics and discussion level
—available expertise to provide pertinent facts
—avoiding conflict-producing agenda items or creating disinterest in these
—satisfaction for each cohort member in influencing the decision

Cohesiveness Through Group Problem Solving. Group problem solving contributes to cohesiveness when cohorts participate in meaningful and effective ways to arrive at group-generated solutions. Conditions contributing to effective cohort group problem solving include:

—heterogeneity of group, i.e., age, sex, etc., differences
—extensive variety of perspectives and alternative solutions entertained
—all cohort members contributing to solution with acknowledged contribution

Cohesiveness Through Mutual Tolerance and Support. Central to group cohesiveness is tolerance and acceptance of individual eccentricities or deviancies, and supportive advocacy on each other's behalf. Indeed, for patients who are to remain out of the hospital, socially supportive living situations are essential.[7] Design criteria for developing cohorts must therefore include the following criteria:[8]

—encouragement of positive regard for each of the other members within cohort
—inculcation in the principle that the only reliable and legitimate means of gaining self-esteem is to grant esteem and respect to others[9]
—assignment of each member of cohort the personal task of learning about each other member, making judgments and receiving feedback about the validity of their judgments[10]

Cohesiveness Through Autonomy of Group Functions. As much as possible, cohort groups need be structured to function independently and autonomously for group cohesiveness. Groups must be given autonomy, with the staff members serving merely as resource persons. Leadership and heterogeneity must also be considered, e.g., groups of individuals who talk the most, versus those who talk the least. Complex tasks are performed poorly by groups where all have high sociability levels. Where groups are composed of all low sociability levels, leadership never emerges.[11] No live-in staff should be provided. The staff should have only minimal contact. Only the assigned lodge staff should be on call, not those of the hospital. A technical consultant is called in when survival in the community becomes a problem. Cohort members control the group decision-making practices. Professional leaders select members, but remain intentionally in the background. The purpose is to increase members' skills in social participation, and to interrupt the vicious cycle of damaged self-esteem.[12]

Implementation of the Cohort Cohesiveness Functions. The basic cohort, and its cohesiveness development, is implemented by documented rules and procedures. Staff skills, in catalyzing the cohesive functions, require special training and development, since such social-psychological techniques are not currently extant for working with a population of chronic mental patients.

Evaluation of Cohesiveness. Situational evaluation is necessary in order to determine if cohesiveness is developing effectively, or if additional measures must be undertaken to further its development. Table 9 presents an evaluation checklist designed to assess cohort cohesiveness in various situations. An impartial observer should complete the checklist at weekly or biweekly intervals, and advise training staff of cohort weaknesses to be improved.

Living-Skills Assessment. Individual patient assessment is needed to determine essential personal deficiencies of each member of the cohort. Competency measures need be sought to identify areas that need enrichment in social, domestic, and community life and total self-image. Patterning of the conventional patient ideal must be deemphasized—quiet, cooperative, punctual in taking medication, etc., where strict conformity, powerlessness and subordination of the patients constitute staff expectancies.[13] Rather, emphasis in indoctrination, discipline, training and evaluation must be shifted to initiative, self-reliance, support of cohort members and self-determining skills.

Table 10 presents a living-skills assessment checklist completed for each patient through observations and patient interviews. The form, with notations and detailed performance items necessary to describe specifics, is completed by staff during the first week to identify skill deficiencies, and subsequent weeks thereafter to serve in progress assessment.

Living-Skills Indoctrination. A living-skills circle of indoctrination is illustrated in Figure 23. Patients enter the circle at any point in one of the four quadrants in progress. The complete circle of quadrants is completed over a 30-day cycle, with each patient receiving the full indoctrination of the circle, over a 30-day length of stay, regardless of the quadrant or area of indoctrination at which he enters. The focus is therewith shifted from a disease or medical model to an educational or training model. The development procedures thus are designed to

TABLE 9
COHORT COHESIVENESS EVALUATION CHECKLIST

Situation	Rating "0 to 5" on Cohesiveness*					
	Decision Making	Problem Solving	Mutual Tolerance	Mutual Support	Autonomy	Staying Together
Informal Unit Living						
Meal Time						
Group Meetings						
Education/Training Classes						
Recreational/Social Functions						
Vocational Training						
Factory Work						
Nonresidential Lodge Functions						
Total Average Score						

* A cohesiveness rating of "0" indicates no interaction; a rating of "5" indicates a tightly knit social interaction of cohorts; an intermediate rating between 0 and 5 indicates somewhere between no interaction and a tightly knit unit, e.g., straggled.

foster the individual patient's growth within the cohort in keeping with his own needs and potentialities.[14]

Development and design problems within this phase include:

1. Conversion and adaptation of staff skills, conventionally oriented to "treatment of mental illness," to individual and group learning processes.
2. Development or application of effective individual and group training or didactic techniques in keeping with sound learning theory and practice, e.g., role playing, psychodrama, etc.

TABLE 10
Individual Living-Skills Assessment Checklist

Patient's Name _____   Date _____   Rater _____

Discharge Target _____

Task and Skill Area (See "Living Skills Code (LSC) Sheet")*	(1) Enter Problem Code(s) from LSC Sheet	(2) Enter "0" to "5" rating (3) Situational Capability			Instructions
		Initial Rating	After Two Weeks	After One Month	Upon Termination
Self-Management					1. Enter codes only for those detailed items requiring attention relative to discharge target.
Control of Aberrant Behaviors					
Control of Health and Diet					2. Enter "0" in columns under "Situational Capability" if no control capability is indicated; "1" if rarely; "2" if sometimes; "3" if usually; "4" if frequently; and "5" if always.
Facial Grooming					
Body Grooming and Dress					
Self-Regard/Image					
Educational Achievement					

* "LSC's" are coded data sheets used by Mental Health Technicians to identify problematic behaviors or task elements in patient performance.

TABLE 10 (CONT'D.)

Domestic Management

Control of Spending

Cooking and Cleaning

Budgetary Planning

Knowledge of
   Helping Agencies

Family or Group
   Living Control

Interpersonal/Social Affairs

Enjoyment of
   Leisure Time

Knowledge of Community

Participation in Recreation

Social Affairs Attendance

Other (Describe)

3. "Situational Capability" refers to the general level at which the patient performs, and the varied circumstances, social and physical settings in which he is able to accomplish the tasks.

NOTE: Various interview, observational and testing sources may be used to obtain rating data. Ratings may be interpretive and judgemental, but should provide a basis in essential skill deficiencies in correction for discharge target.

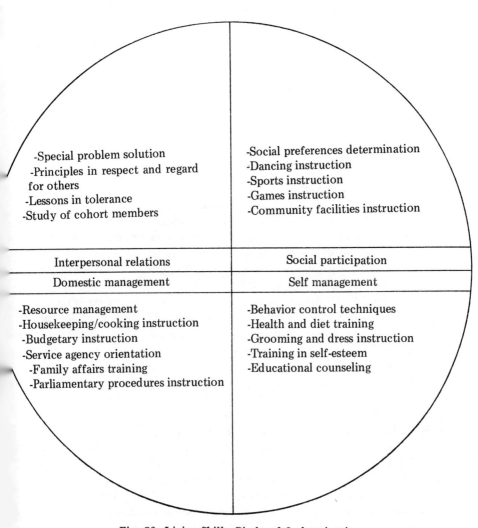

Fig. 23. Living-Skills Circle of Indoctrination

3. Allocation of appropriate and effective training or didactic functions to machine or teaching devices, or to staff-directed teaching functions. This includes the use of training films, teaching machines, video and audio tapes, etc.

All phases of indoctrination must be oriented to improved community living, and, specifically, to work and adjustment in the non-residential lodge environment.

Community Placement. Community placement of patients is a subsidiary function, though essential in the system. Patients are slated for residence in group-care homes or private living accommodations, with public-aid support, or support by their wages through productive employment. Advanced planning and implementation criteria include:

—arrangements for easy access to lodge during daytime and evening hours
—arrangements for communication among cohort members in the community
—arrangements for compatible living situations within or outside community in which patient originated

Phase C. *Indoctrination in Vocational Skills.* The vocational phase of the mission includes vocational skills assessment, vocational-skills indoctrination, and vocational and job placement liaison. Skills assessment functions must identify essential deficiencies for indoctrination, correction, or improvement. Indoctrination functions must include periodic feedback provisions in progressively focusing on areas for behavior and attitude adjustment.[15]

Vocational Skills Assessment. Individual patient assessment is necessary to assist in reinforcing areas of vocational skills and attitudes for further training, or for acquiring and holding a job. Table 11 presents an individual assessment checklist for bracketing areas of vocational skills deficiency. Contact with former employers may further indicate specific areas of deficiency. Ratings may be completed prior to factory participation, and weekly, during the factory work activity. Accompanying notations or detailed work items may serve to pin point the skills deficiencies for correction or improvement.

Vocational Skills Indoctrination. Essential realism must be applied in this phase of the mission to reinforce the patients orientation in the workaday world. Phenothiazines and other drugs for patient management should be avoided to maintain cognitive sensitivity and to permit elements of the work experience to operate unimpededly in improving the patient's vocational outlook. Criteria for design of vocational-skills indoctrination include the following:

—anticipation of vocational and adjustment needs
—training related to why patient was discharged from previous job
—paid wages
—permissiveness, with graduated work element demands
—confidence engendering
—requirements for instructions, work routines, punctuality, quality and quantity of work

TABLE 11

√	Rate 0 to 5 on Proficiency**			
	Initiative*	Self Reliance	Performance Reliability	Cohort Contribution
On time for work				
Communicating on job				
Following instructions				
Getting on with other workers				
Skill with hands				
Use of tools				
Attention to work				
Speed of work				
Quality of work				
Judgment on the job				
Reaction to supervisors				
Attitude toward work				
Performance on complex jobs				
Adaptability to new jobs				
Acceptance of routine				
Other (describe) :				

√ Check those areas which previous employer found deficient or a reason for discharge.

* "Initiative"—seeks or begins action himself.
"Self-Reliance"—can be relied upon to participate regardless of situation or circumstance.
"Performance Reliability"—actions consistently carried through to completion.
"Cohort Contribution"—helps other members of cohort.

** A proficiency rating of "0" indicates no capability whatsoever; a rating of "5" indicates high-level performance and growth; an intermediate rating from "0" to "5" indicates somewhere between no capability and high-level performance, e.g., an entry of "3" under "Initiative" in being "on time for work" would mean that he is sometimes on time without being prodded.

—use of time clocks

—realism in objective supervision

—explicit work standards

—feedback to patients on job performance elements

—rewards to patients for all positive elements of work performance

—participation by outside mercantilists for sponsorship and employment contacts

Vocational Liaison. Liaison with community facilities is necessary to promote the full employment potential of the patients after discharge. Vocational training in the community may be obtained through the Division of Vocational Rehabilitation, while the State Employment Service, and direct contacts with commerce and industry, may afford necessary job opportunities for employment.

Phase D. *Mutualizing Cohort Support in the Community.* The final phase of the rehabilitation mission is in establishing the operation of mutually supportive chronic-patient cohorts in the community. This phase consists of facilities planning and preparation, development of in-community cohort cohesiveness, and the maintenance and assessment of mutuality of support among cohorts in the community.

Facilities Planning and Preparation. Facilities planning for cohort community interaction must dovetail with residential and job placement and activity functions. The following criteria must thus be considered in facilities planning:

—geographic location readily accessible to all cohort members, by foot, bus or car, at all operating periods

—hours of facilities operation compatible with work and other outside activities of each cohort member to insure cohort interaction

—requirements for building, fix-up, paint-up, etc., or service to the community compatible with cohort skills, capability and interest

—recreational equipment, games and activities-planning compatible with practical community social interaction, e.g., no ceramics

—provisions for cohort contacts with no-show cohort members

—provisions for emergency support of cohort members in problems of personal crisis

—a scheduled, dated plan for Meyer staff withdrawal and mental health center phase-in of responsibility

Operation of In-Community Cohort Cohesiveness. In the transition of group cohesive functions from the hospital to a community situation, essential differences in the milieu must be considered. Also, possible extinction effects from delays in establishing cohesive linkages in the community must also be avoided, and/or compensating reinforcement must be provided. The following criteria are indicated for operation of the community club facility for cohort cohesiveness:

ORGANIZATIONAL SETUP

—Patients discharged as a cohort with social cohesiveness
—Entire group oriented, able and willing to perform useful community service projects
—Club oriented to life-long affiliation tightness of membership, or until individual patient is completely stabilized in the community
—Essential cohort membership continues to participate in decision-making and problem solving
—A quiet room is made available
—A room large enough for meeting provided
—Facility design, warm and cheerful atmosphere; seating arranged to encourage two, three or four people to sit together
—Club oriented toward health, and enlarging number of concerned friends, developing skills and interests
—Policy enforced that any unhurtful behavior acceptable. One does as he pleases while in the lodge, but in public must conform
—Clubs run along parliamentary lines; officers elected, dues collected, and projects planned. Professionals attend, but remain in background
—Members pay own way
—Patients participate in planning programs, playing games, serving on committees, etc.
—Meetings held at least once a week. More frequent meetings encouraged
—New and meaningful social roles created. Achievement dependent upon each other for survival
—Follow-up to be made by cohort members to determine why other members are absent
—Residents themselves serve to cook, wash dishes and act as bookkeepers, business or residential managers, and in medication distribution

RECREATIONAL ACTIVITIES

—Group singing, parties and dances included
—Minimal skills required, such as badminton, miniature golf and bowling
—Outdoor sports to be included, such as volleyball, badminton, etc.

—Occupational therapy avoided. Sewing, knitting, hair styling, painting, budgeting, woodworking, furniture restoration and sedentary games, etc., provided to create further interest in leisure time

—Activities include structured games and trips

—Passive activities avoided. Picnics, games etc., requiring cooperation and teamwork encouraged

—Normal activities, also found in community, encouraged, such as card playing, swimming, etc.; institutional activities avoided, such as basketry or weaving. Childish games avoided

DISCUSSION GROUPS

—Group discussions to include total membership and small groups

—Problem-solving groups formed for role playing, group discussions and creative writing

—Insight therapy avoided, with coping therapy highlighted

PROFESSIONAL STAFF ROLES

—Initially, a professional initiates group functions and coordinates work; all functions then assumed by the patients

—Only assigned lodge staff on call. A technical consultant called in when survival in community becomes a problem. Lodge members control the group decision-making practices

—Volunteers employed who are trained to deal with mentally ill people in activity therapy and other skills[16]

Maintenance and Assessment of Cohort Community Functions. In sustaining a continuing supportive operation among discharged cohort, the basic sociopsychological principle of symbiosis must be maintained. Parasitic or one-sided, noncontributive relationships tend to break down. Therefore, it becomes essential to sustaining the cohesive function that each member becomes a mutually supporting component of the cohort.

Table 12 presents a checklist for rating various aspects of mutual cohort support in various situations and on the dimensions of initiative, depth, consistency and symbiosis. Cohorts, within the larger lodge grouping, must of course be identified for rating; nor should they be considered to be rigidly defined as impermeable, closed groups of the original five formed in the hospital. Rather, other more natural cohort groupings emerging within the club may constitute the more meaningful unit for assessment.

TABLE 12
COHORT MUTUAL-SUPPORT EVALUATION CHECKLIST

Situation	Rating "0" to "5" on Mutual Support**			
	Initiative*	Depth	Consistency	Symbiosis
Member sick or absent				
Member disturbed/abusive when in cohort				
Member declines to participate in cohort activity				
Member has trouble with family or living environment				
Member has trouble with work environment				
Member has trouble in public with his behavior problems				
Other (please describe):				

* "Initiative"–cohort member (s) begin supportive action themselves
"Depth"–length of support to which cohort member (s) extend themselves
"Consistency"–persistency with which cohort support is provided regardless of how often need arises.
"Symbiosis"–Extent to which *all* members contribute
** A mutual-support rating of "0" indicates that no member of the group makes an effort or registers concern; a rating of "5" indicates several cohort members become involved and all register concern; intermediate ratings indicate somewhere between absolutely no support and total concern.

The ratings may be best completed each week during the first month and each month thereafter, and serve as feedback or monitoring data for staff catalytic action.

SYSTEM EVALUATION

Success of the entire rehabilitation system is predicated upon assessment of projected outcome goals. These must be correlated with specific functions, and determinations made as to the extent to which subsystem functions are contributing to each objective. Continual administrative monitoring for feedback is practiced, to correct or eliminate ineffective components, or improve functional areas that are only marginally effective. Constant improvement is sought in measurable outcome parameters, including the following:

1. Percent of daily lodge attendance by the total number discharged to community as cohorts
2. Cohort cohesiveness in lodge:
   —percent actively participating in (a) group meetings, and (b) group recreation
   —percent actively supporting other lodge members in daily life situations
   —number of productive community projects completed per month
   —ratio of lodge-member to staff-member participation in managing the lodge
3. Percent requiring reinstitutionalization within a one-year period
4. Percent entering job training programs within a one-year period
5. Percent sustained in productive job employment over a one-year period

Evaluative criteria of subsystem performance contributing to these total system measures include:

Cohort Gathering. Lodge attendance and cohort cohesiveness may hinge about selection criteria for cohorts. Geographic area of origin, site of cohort gathering, etc., may prove to be central to these lodge functions and require constant feedback for improvement.

Living Skills Indoctrination. Active participation and mutual support by lodge members in the later community phase may become largely a function of how effectively the groups are welded together in this earlier phase; or, on the other hand, no measure of cohesiveness in this subsystem may bear on total system measures of success in rehabilitation. Ongoing feedback measures are needed to determine this, and to assess, adjust and improve performance in the subsytem.

Vocational Skills Indoctrination. The extent to which the vocational-skill system measures (4 and 5 above) are effected by this subsystem, also requires close monitoring.

Feedback must be obtained on failure of cohort members to enter job training programs, to obtain jobs and be sustained in productive employment, in order to improve performance elements of this subsystem.

BUDGETARY SUSTENANCE

In order to sustain operation of the rehabilitation system, essential budgetary support must be assured. State funding continues to be the primary source. The vocational-skills acquisition subsystem or factory is supported by an annual department allocation, though it is intended that the factory operation become wholly or partially self-supporting.

With continuing marketing success, increased income from sales and accrual of resources is partially diverted to increasing the salaries paid to the factory workers.

Financial support of the community lodge is also provided by the state Department of Mental Health. Administration, at the outset, is provided by state-employed staff. It is expected that administrative functions, however, will be eventually assumed by grant-in-aid agencies through continuing state funding.

A FINAL SYSTEMS NOTE

It is important to realize that, for a successful systems operation, rehabilitation is contingent upon effective performance of all system elements. Breakdown of any of the essential components in cohort formation, cohesiveness development, vocational skills indoctrination, community support, etc., may result in breakdown of the total system, i.e., measures of outcome as described above will be achievable on an otherwise only random basis. Administratively this means that a total systems discipline must be maintained throughout the life of the system.

NOTES TO CHAPTER 12

1. For further discussion of opened and closed systems, refer to L. von Bertalanffy, "The Theory of Open Systems in Physics and Biology," *Science* 3 (1950) :23–28; and R. Nelson and J. Burgess, "An Open Adaptive Systems Analysis of Community Mental Health Services," *Social Psychiatry* 8, no. 4 (1973) :192–97.

2. A. Hare, *Handbook of Small Groups* (New York: The Free Press, 1962) ; E. Miller, *Systems of Organization* (London: Tavistock, 1967) ; J. Hackman, "Group Size and Task Type," *Sociometry* 33 (1970) ; J. Mann, *Changing Human Behavior* (New York: Charles Scribner's Sons, 1965) .

3. G. Fairweather et al., *Community Life for the Mentally Ill: An Alternative to Institutional Care* (Chicago: Aldine Company, 1969) .

4. L. McDonald and G. Gregory, "The Fort Logan Lodge. Intentional Community for Chronic Mental Patients," U.S. Public Health Grant, No. 1 RO1 MH15853-02 (Fort Logan Mental Health Center, Denver, Colorado, 1971) ; D. Sanders, "Innovative Environments in the Community: A Life for the Chronic Patient," *Schizophrenic Bulletin*, Fall 1972, pp. 49 ff.

5. M. Palmer, "The Social Club. A Bridge from Mental Hospital to Community" (National Association for Mental Health, 10 Columbus Circle, New York, New York 10019, 1966) .

6. W. Vinacke, and W. Wilson, *Dimensions of Social Psychology* (Glenview, Ill.: Scott, Foresman and Company, 1964).

7. Sanders, "Innovative Environments. . . ."

8. J. Martin, *The Tolerant Personality* (Detroit, Mich.: Wayne State University Press, 1964); Palmer, "The Social Club. . . ."; H. Smith, *Sensitivity to People* (New York: McGraw-Hill Book Company, 1966); C. Kiesler et al., *Attitude Change: A Critical Analysis of Theoretical Approaches* (New York: John Wiley & Sons, Inc., 1969).

9. W. Rosengren, "The Self in the Emotionally Disturbed," *American Journal of Sociology* 56 (March 1961) :454 ff.; S. Coppersmith, *The Antecedents of Self-Esteem* (San Francisco, Calif.: W. H. Freeman, 1967); A. Fontanna et al., "Presentation of Self in Mental Illness," *Journal of Consulting and Clinical Psychology* 32 (1968) :110 ff.; B. Grossman, "Enhancing the Self," *Exceptional Children* 38 (November 1971) :248 ff.; H. Weissman et al., "Changes in Self-Regard, Creativity and Interpersonal Behavior as a Function of Audio-tape Encounter Group Experiences," *Psychological Reports* 31 (1972) :975 ff.; R. Hedges, "Photography and Self-Concept," *Audiovisual Instruction* 17 (May 1972) :26 ff.

10. H. Smith, *Sensitivity to People*; C. Kiesler, *Attitude Change*.

11. G. Fairweather, *Community Life for the Mentally Ill*.

12. I. Bierer, "A New Form of Group Psychotherapy," *Mental Health* 5, no. 22 (1944) :1363 ff.; P. Hanson et al., "Autonomous Groups in Human Relations Training for Psychiatric Patients," *The Journal of Applied Behavioral Science* 2 (1966) :305 ff.

13. M. Wiernasz, "Quarterway House Program for the Hospitalized Mentally Ill," *Social Work,* November 1972, pp. 72 ff.

14. I. Babow, "An Instrument for Studying Leisure Activities of Mental Patients" (Rehabilitation Services, Napa State Hospital, Imola, California, 1968).

15. W. Wadsworth et al., *The Need to Work* (Philadelphia: Smith, Kline and French, 1962); W. Query, *The Hospital Factory in Rehabilitation* (San Francisco, Calif.: Jossey-Bass, Inc., 1968); M. Barbee, "Relationship of Work to Length of Stay and Readmission," *Journal of Consulting and Clinical Psychology* 33 (1969) : 735 ff.; B. Black, *Principles of Industrial Therapy for the Mentally Ill* (New York: Grune and Stratton, Inc., 1970); J. Kunce, "Is Work Therapy Therapeutic," *Rehabilitation Literature* 31 (1970) :297 ff.; R. Walker et al., "Occupation Therapy Patients to Paid Work," *Rehabilitation Literature* 32 (1971) :360 ff.

16. A. Bill, "Social Clubs Help Prevent Readmissions," *Hospitals and Community Psychiatry* 21, no. 5 (1970) :41 ff.

13

Human-Service Systems
with Problems of Fragmentation

The majority of human-service subsystems in practice largely seem to be fragmented in design and operation. Coded departments of government and public service, for example, all appear to operate in isolation as though existing as independent entities. Cooperation among such departments as Public Aid, Public Health, and Mental Health, when it does operate, seems forced and strained, and raked with inefficiencies.[1]

The domiciliary subsystem must involve multiple interacting elements among other subsystems (e.g., surface traffic, fire control, sanitation, etc.), as well as among components within the subsystem itself. Likewise, commerce and industry, the criminal justice subsystem, health, mental health and public welfare subsystems require integrated interactive design of component elements. The design of the interactive elements of such human-service subsystems, however, most frequently is wholly neglected, or considered on only a casual or intuitive basis. Each subsystem, in fact, presents formidable barriers to such integration, when the primary criterion for low-cost housing is just that; or the uninformed, unorganized consumer accepts housing that cannot be maintained, is inaccessible to work and commerce, presents high risk of fatality in fire hazards, etc. When administrative authority and social design goals for habitability, corrections, rehabilitation, etc., are ill defined, indeed a mish-mash of outcome operations may well be inevitable.

The purpose of the present chapter is to address several subsystems not readily lending themselves to integrated analysis, and to expound on several possible methods in a human-factors approach that have been applied for the improvement of fragmented subsystems.

DOMICILIARY AND COMMERCIAL SUBSYSTEMS

Policy, development goals, and specifications are usually not extensively articulated by builders and developers of housing and commercial projects on perhaps other than completion dates, occupancy limitations, materials, costs, etc. System goals, however, would indeed seem most important to project success. For example, in a housing renewal program the population characteristics, their socioeconomic and rural-urban background, etc., would seem to be an essential consideration for design, training, and regulation of occupant practices.[2] In addition to the more total system problems of health, economic support, recreation, etc., the internal subsystem problems of ingress and egress to individual housing units, safety (e.g., fire, criminal assault, etc.), maintainability and cyclic renewal, etc., need also be considered. These, however, are most frequently neglected in design; nor is a comprehensive overview of user behavior, space layout, and user's preferences incorporated in design. Current environmental designers have developed methods for probing interior design and layout for dining facilities, bachelor and family housing, offices, stores, educational facilities, wards and clinics, etc. Figure 24 outlines a general method of analysis applied to these various domiciliary and commercial subsystems. Dependent variables become essentially the measures of irritations, annoyances, or complaints (for statistical purposes, usually measured as a function of a semantic-differential function). The objective determinants are then obtained in terms of spatial area, defective or inferior equipment, etc. These are then statistically compared with those of low-complaint units to obtain standards for design criteria. Human engineering standards and task data may also be employed in the analysis. Table 13 further outlines some representative measures obtained for a commercial establishment. Measures may be derived from available instruments, or developed with nominal (e.g., odor), ordinal (e.g., predominance of color), interval (e.g., attitude measures), or ratio scales (e.g., number per square foot area).

DOMICILES AND COMMUNITY PROBLEMS

Housing and commercial building construction may often be predicated on outmoded design standards, while new and sometimes alarming problems arise in conjunction with other subsystem variables, e.g., factory air pollution, noise disturbance, etc.

The French architect François Blondel, 300 years ago, measured the human stride at about 25 inches average. Over the centuries, while

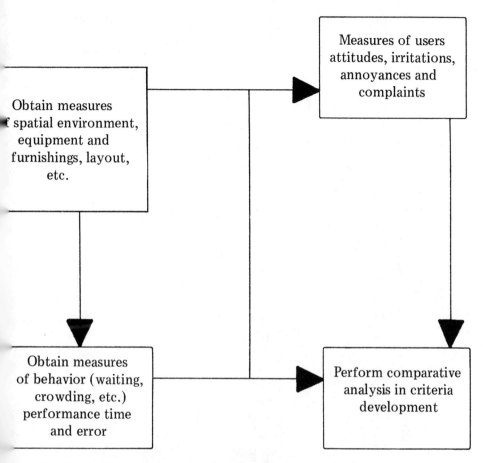

Fig. 24. A Method for Interior Environmental Analyses and Evaluation

statistical body sizes have continued to change with each generation, 25 inches has remained the measure about which the standard stair step and building codes are designed. Today such a standard becomes a nuisance; often staircases become a source of hazard, precipitating human error in ascending or descending behavior.

THE NOISE POLLUTION PROBLEM

Noise pollution has become both a nuisance and hazard to community domiciles.[3] In recent years, an increasing number of studies have been directed to nuisance or annoyance conditions of noise, gaining its chief impetus from sonic boom studies during the 1960s

TABLE 13

AVAILABLE MEASURES IN INTERIOR-DESIGN ANALYSIS

Attitude or Behavior Measures	Objective Measures of Environment	Comparative Analyses
Semantic Differential	Number per square foot area	Statistical comparison of high-complaint units with low-complaint units
—crowded	Aisle widths	
—fast	Number, predominance and variety of color	
—colorful	Flow rate per square foot area	
—efficient	Light measures	
—organized	Descriptive odor sources	Evaluation against human engineering design standards
—noise	Decibel measures	
—odor	Layout	
—clean	Equipment features	
—light		
Time Measures		
—waiting		
—flow path		
—operational performance		
Error Measures		
Items missed		
Breakage rates		
Over-under charges		

when the Supersonic Transport was under development and had threatened to become a major community nuisance. Early psychological measures concerned loudness estimates against magnitude and frequency (now called "hertz" or hz) measures. Examples of early rating scales include:

Overall Sound Pressure Level (OASPL)
A-weighted Sound Level [L_A in db (A)]
Loudness Level (LL in phons)
Articulation Index (AI in percent)
Speech Interference Level (SIL)

Rating procedures have been further developed to include the context of the noise and peculiar characteristics of duration, impulsiveness or intermittency, and the community situation the noise penetrates, including background noise; neighborhood; time of day, week, or year. Recently derived community noise measures include:

Perceived Noise Level (PNL in PNdb)
Noise and Number Index (NNI in amplitude)
Traffic Noise Index (TNI in amplitude)
Noise Pollution Level (NPL in amplitude)
Composite Noise Rating (CNR)
Noise Exposure Forecast (NEF)

Housing site location criteria are currently based on Noise Exposure Forecast contours. This measure has recently been adopted by the U. S. Department of Transportation. Contours take into account such operations as total number of aircraft operations (or other noise-generating sources), various types of aircraft operations, proportion of daytime to nighttime operations off various sites, etc., weighted to accord with responses of various community dwellers reacting to the noise.

Noise is, of course, generated from multiple sources about community domiciles, e.g., surface automobile and truck traffic, factory operations, military exercises, etc. Figure 25 illustrates common acceptance standards of road noise as a function of truck-road noise.[4] Those attributable to aircraft operations are based on random population samples about blocks forward of nominal noise exposure zones and the public annoyance reaction.[5]

Table 14 shows the number of take-offs and landings for several airports, and the number of persons per square mile about the airports. Figure 26 shows that, in plotting number of flight operations

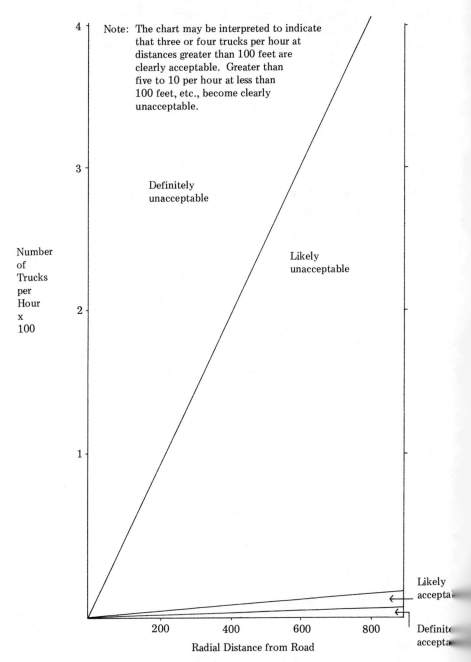

Fig. 25. Acceptable Standards in Domiciles for Truck Road Noise

against public survey data showing percent of the population annoyed, it tends to indicate no appreciable direct relationship. However, plotting the number of flight operations against population density, and population density against percent annoyed (Figures 27 and 28), does indicate a relationship. The relationship between percent annoyed and overt complaints is plotted in Figure 29.[6] Population per square mile tends to assume a linear relationship with number of take-offs and landings per year. Population per square mile in each Standard Metropolitan Statistical Area (SMSA) also appears to be related in a linear fashion to percent of highly annoyed population.

TABLE 14

FLIGHT-OPERATIONS SAMPLE IN THE UNITED STATES

SMSA* Population	Annual total no. of take-offs & landings**	Sq. miles	SMSA population per sq. mile
Reno, Nevada 72,836	157,774	304	239
Denver, Colo. 514,678	400,000	167.6	3070
Miami, Fla. 350,000	299,631	71	4577
Chicago, Ill. 4,170,400	750,000	288	14479
Lawton, Okla. 74,470	3,000	44.2	1684

* SMSA—Standard Metropolitan Statistical Area: total geographic area designated by the Census Bureau as such when having a city of at least 50,000 population within its boundaries, and more than 250,000 total population.
** Obtained from flight operations towers or traffic control offices at respective airports during June of 1975.

Though no clear-cut principles are yet to emerge from such studies, the implications are that aircraft and other noise-generating sources are annoying to community domiciles. A future solution might lie in removing airports and other major noise sources to more distant and remote areas of lesser population density, i.e., of course, with rapid access by surface or subsurface transportation (see Figure 30).

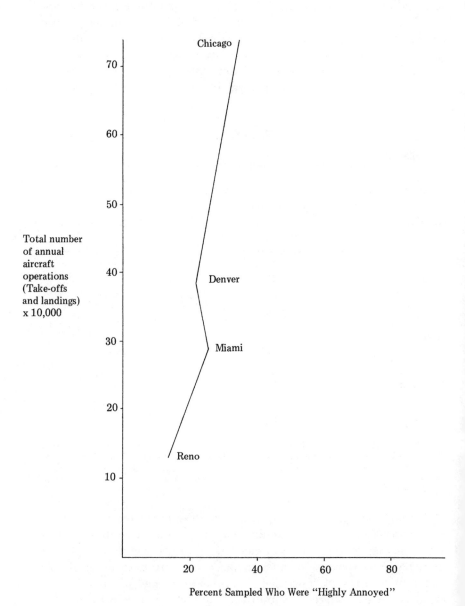

Fig. 26. Total Number of All Aircraft Operations in the
SMSA Relative to Percent Annoyed

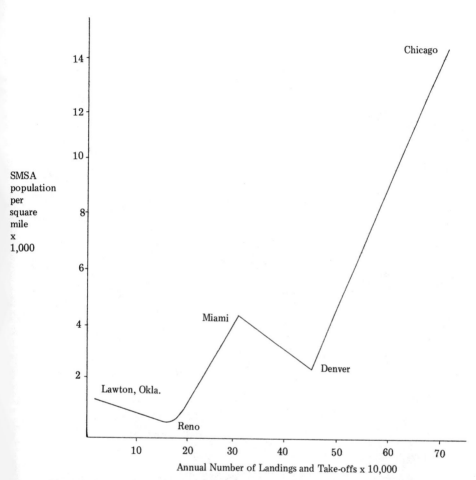

Fig. 27. Number of Aircraft Operations per Square Mile of Population

The domiciliary and commercial subsystems, of course, interact with the more total human-service system, in traffic control, sanitation, health, etc. However, such methods as described above provide a basis for situational and operational improvements, even though the subsystems are taken independently, each with an isolated, fragmented character of operation.

CRIMINAL JUSTICE SUBSYSTEM

The criminal justice subsystem may be seen as a highly fragmented

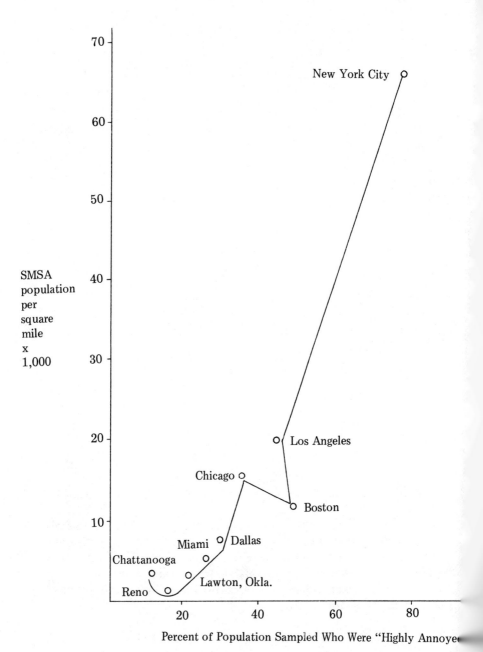

Fig. 28. Population Density and Percent Annoyance with Aircraft Noise

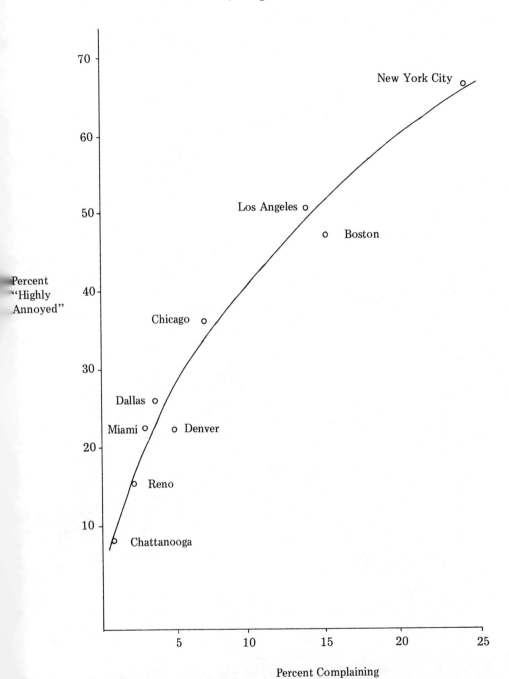

Fig. 29. Empirically Derived Complaint Frequency as a Function of
Those Found "Highly Annoyed" in the Survey

Scale: 1 inch = ca. 1 mile

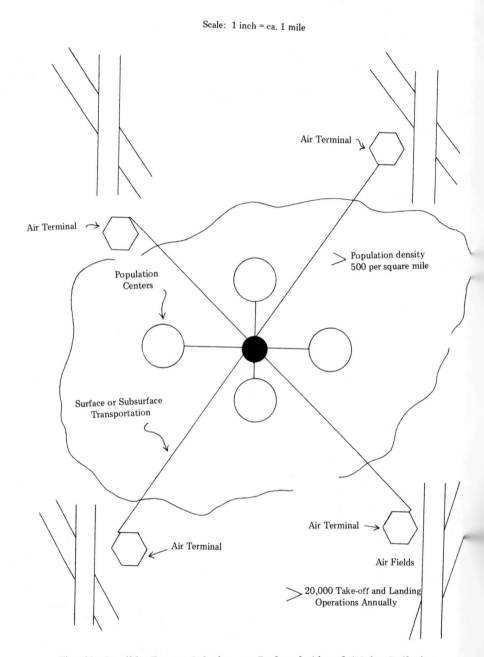

Fig. 30. Possible Future Solution to Reduced Aircraft/Noise Pollution

subsystem, lacking both a definitive policy stipulation and a goal-directed design orientation.

In the United States, for example, juvenile crime has risen by 1,600 percent over the past twenty years. More crimes have been committed by children under 15 than by adults over 25. Estimates indicate that over half of all crimes are committed by juveniles. In a recent year, police arrested 2½ million under 18. In Los Angeles, juveniles are responsible for more than a third of major crimes; in Phoenix, Arizona, estimates are 80 percent, and in New York City, burglary and rape charges against juveniles have doubled in four years.

Juvenile offenders are often emotionally disturbed children whose home life is shattered; reasons for high juvenile crime rates are primarily system-borne—poverty and overcrowding, city slums, high rates of unemployment, the easy access to guns and knives, etc.

It is in such an endless fragmented cycle of corrections that juvenile criminals become adult professional offenders. Martinson, in his criminology studies, concludes that all available criminal rehabilitative efforts reported in English from 1945 through 1967 have had no appreciable effect on recidivism.[7] Overall conclusions suggest that the practice of forced state treatment is based more on myth than scientific evidence. Prison "treatment" programs simply cannot substitute for normal society, nor compensate for the damage done by removing offenders from social growth and opportunities. If corrections do not correct, holding young offenders for long periods of time in reformatories under indeterminate sentence is futile. Sums of money now used in enriching prison programs may better be employed for probation and parole functions, more effective police patrol, crime prevention programs in boys' clubs, etc. Rehabilitation methods may better involve small probation and parole case loads, intensive supervision, early release from confinement, varying degrees of custody, casework, individual counseling, psychotherapy, group therapies, milieu therapy, halfway houses, pre-release guidance centers, tranquilizing drugs, plastic surgery, and general human services. For any measure of rehabilitation to be accomplished, indeed, the incarcerated population would require more extensive detailed analysis of objective needs. In fact, a paradox appears to operate, where a retributive or punitive philosophy and practice dominates the subsystem, often obscuring and overriding rehabilitative goals. Whereas if the basic problem of a social deviant or psychopath was one of nonparticipation or nonlearning of the social code, the punitive practices reinforce such nonparticipation through isolation and prolonged nonsocial participation. As a most severely fragmented subsystem, new and innovative methods are needed perhaps in first clarifying and expanding civil rights functions for broader social participation leading more emphatically in the direction of socialization (in the middle-class, society-at-large sense) . Then it may become more feasible to adapt and design methods that will effectively impact on this most severely fragmented but, above all, an inaccessible subsystem.

HEALTH SUBSYSTEM

The health-care subsystem may be considered to consist of preventive, emergency, diagnostic, therapeutic, and rehabilitative functional components. These include primary medical care and family medicine. Preventive functions may consist of various health education, disease control, early detection, and information programs on symptom recognition and health resources. Emergency functions are those of a fast reaction to trauma, or the sudden onset of a disease or injury, such as the system described in chapter 11. Diagnostic functions include varying degrees of sophistication in the use of methods and instruments for determination of specific diseases, organic, physical, emotional, and social malfunctions. Therapy then consists of various treatment and corrective measures and applications, while rehabilitation is the procedure necessary to bring about full recovery to a "normal" status.

The various sites at which such functions are carried out include offices of solo practitioners; outpatient dispensaries; voluntary, private, or public clinics (health department clinics; group practice clinics or those in factories or schools) ; the home of the patient; or in hospital wards. Patients may be in the vertical or horizontal posture in the execution of these functions. Methods of funding may be private, public, or by voluntary or involuntary insurance coverage. Compensation of individual physicians or paramedical personnel may be by fee in private treatment or sessional payment, or by salary.

These functional components of a health-care subsystem are, again, generally fragmented in design and operation, though naturally interactive in the implications of outcome. Introduction of the Salk vaccine for poliomyelitis, for example, had major impact in reducing therapeutic and rehabilitation functions, as did preventive functions in tuberculosis. Though each functional component of the health-care subsystem tends generally to be isolated without effective planning and allocation of resources, each area does indeed lend itself to improvement through separate system studies and applications.

APPLICATIONS IN THE AREA OF PREVENTION

Preventive measures in health must relate to desirable and operationally determined states of health.[8] Reasonably accurate health measures for preventive programs may be often obtained from public health departments on reportable diseases.[9] In studying cause-of-death data, for example, childhood mortality rates are largely attributable to nonmotor vehicle accidents and pneumonia, whereas adolescent and young adult death rates are largely accounted for by automobile accidents. Syphilis and gonorrhea are also highest in the latter population. In fact, mortality and incidence of reportable diseases are higher by a factor of two to five for blacks and other poverty groups. Such epidemiological data, when available or derived, provide the basis for preventive program design.[10] This might take the form of educational and informational dispersion to the high-risk groups (cf., venereal disease with young adults), antibody generation programs, preventive medication, etc. The human-factors contribution in the area of prevention might concern such activities as information display mode and design, developing road and automobile safety design features, design of inoculation instruments to relieve stress and trepidation in patients and to improve operating efficiency of the care givers, etc. Other forms of human-engineering contributions to prevention might concern display problems relative to organic reactions. For example, flicker frequency in displays approaching the Alpha rhythm of the brain (approximately 12 cycles per second) is known to induce epilepsy in organically prone individuals. As a preventive measure, such flicker may thus best be avoided in displays or ornamental arrangements. A human-factors analysis may also be of value in other such contexts where environmental display settings might induce vertigo, nausea, or otherwise reinforce sick roles.

EMERGENCY HEALTH CARE

Human-factors implications in design of a Trauma Center operation were described in chapter 11. Other, more fragmented aspects of emergency-care component design, might also concern emergency room layout, efficient individual and team procedures development, equipment design and personnel emergency training functions.

Stanfield[11] has delineated several human-engineering requirements in the design of a cardiac defibrillator. The reaction time requirements when ventricular fibrillation occurs are fairly clearly indicated. If defibrillation is delayed up to 12 minutes, for example, chances of recovery are less than one percent. The defibrillation equipment must therefore permit an accurate and rapid determination of a fibrillation problem. The application of a 300-watt-seconds electrical power source at the cardiac region must thus be assured within minutes. Human-engineering standards should be addressed to the rapid and reliable determination of fibrillation, control and efficient application through the defribrillator paddles of required current and voltage, proper warning labels and interlocks to assure safety to operating personnel and prevent further aggravated injury or ineffectual applications to the patient. Though in itself often treated as an isolated and fragmented area, the emergency treatment component nicely lends itself to human-engineering design, and the specification of fairly precise parameters and constraints.

HUMAN ENGINEERING OF CONVENTIONAL DIAGNOSTIC INSTRUMENTS

Another most severely fragmented area of medical care perhaps lies in the realm of private practice, and the private practitioner who ministers to the major segment of health-care needs. His procedures, routines, instruments, and training have been largely individualized, with only a modicum of review and critique by peers. Perhaps a major potential for more immediate human-factors contribution lies in the human engineering of the multiple and variegated medical-surgical instruments at his disposal. Adapting spinoffs from many of the sophisticated aeromedical physiological monitoring equipment components and devices, developed by human-factors personnel within recent years in aerospace applications, may become a reasonable goal for civil health care. Constraints operating in the public-private sector, however, of course impose severe limitations on such developments that utilize telemetry, computers, and sophisticated electronic devices. These aerospace developments, however, do provide the potential background and state-of-the-art amenable to a more widespread application in the public and private health sector.

Several classical diagnostic instruments amenable to human-factors analysis are illustrated in Figure 31. The ophthalmoscope and otoscope are visual inspection instruments of eye (retinal) and ear structures for detection of swelling, inflammation, or disease. Specific functions of these instruments do not appear to have been scrutinized in human-factors analysis, nor have the manual-visual operations and dimensions of these instruments been evaluated.

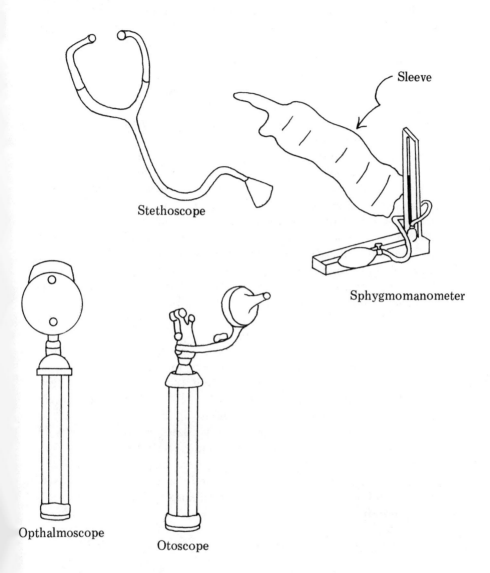

Fig. 31. Several Classical Medical Instruments and Functions
Amenable to Human-Factors Analysis

The sphygmomanometer, used for measuring blood pressure, has also been effectively the same for many decades. The measure consists of collapsing the brachial artery in the arm by pressurizing the wrap-around air-inflatable sleeve with 200–250 mm of mercury or air for tube and gauge pressure reading. As the pressure is dropped at an approximate rate of 5 mm per second, the brachial artery is sounded with the stethoscope—when first heard, the systolic or pumping pressure is noted, and, when fading, the diastolic or resting pressure is recorded. While blood pressure is considered a critically significant measure of heart condition, this classical means of obtaining it is often without a great deal of significance, even in routine physical examinations. It provides no more than a "snapshot" type of datum while the patient is usually seated. The momentary measure is sensitive to both the apprehension of the patient and gross reading error of the recorder, with virtually no variable stresses possible for testing. Cost of the present measuring instruments is less than $40, but growing state-of-the-art in electronic equipment could well mean vastly improved and minimally costly heart-measuring equipment. The computer-processed electrocardiograms developed by Dr. R. Zitnik[12] provide such a promising application which could be made available to private physicians. A portable cart or field unit is employed which consists of an electrical harness for attachment at the patient's cardiac region, a terminal, telephone jack, and a cathode ray tube. The functional procedure is to attach the harness to the patient, plug the terminal unit into the computer through the phone-dialing system, have the patient engage in various "heart-test" activities, and obtain discrete readings and/or interpretations by the computer for the cardiogram and vector-cardiagram (cross-related electrical activity of the heart). Such developments as these may readily lend themselves to standardization and adaptability to a general population of patients and users, e.g., anthropometric sizing and "snap-on" adjustments of harness, optimal layout, access, and control-display design and information processing, etc.

HUMAN ENGINEERING OF MEDICAL SITES

An area that is receiving increasing human-factors attention is the layout and design of work stations, wards, spaces, and equipment in the health-care subsystem. Objectives must include environmental design that will optimally reduce the sick role and the stress from lack of privacy and "being done something with" on the patient, as well as improved staff efficiency through optimal layout and equipment design.[13] Layout and human engineering of medical sites may, on the one hand, require more definitive goals for patient outcome and clarification of patient psychology in a hospital or clinical setting; on the other hand, layout and equipment design for staff operation may best be predicated on classical functions and task analysis procedures, whether these are carried out in general hospitals or medical clinics, or in the office of the private practicing physician.

PRIMARY MEDICAL AND FAMILY HEALTH CARE

Perhaps the most crucially fragmented aspect of the health-care subsystem is in the meaningful and effective delivery of health services to those population segments most critically in need. Indeed, health, as a human-service product, impacts on a wide band of human needs and accomplishments. In the absence of effective health care, unemployment and welfare needs are multiplied and human energies essential for independence and creative endeavors are markedly attenuated.[14] The Health Maintenance Organization Act of 1973 holds promise of expanding health care to a greater number of needy segments in the population, though those most severely neglected may continue to be lacking in essential services.[15]

Survey techniques may become essential in establishing health-care needs for the isolated, critical segments of population currently not served. A technique for accomplishing such surveys was developed in a metropolitan community of central Illinois during February and March of 1971 under the aegis of several cooperating civic groups including mental health. Actual survey work was accomplished by the consumers themselves—a group of organized citizens who called themselves the "Torrence Park Citizens' Committee." The health needs survey was made in the Torrence Park area of the city of Decatur, Illinois, where 70 percent of household incomes was less than $5,000, to determine where critical deficiencies existed in the delivery of individual health-care services. A survey was prepared jointly by members of the Torrence Park Citizens' Committee, the South Central Illinois Regional Health Planning Council, the Decatur Macon County Office of Economic Opportunity, and the Department of Mental Health. Approximately ten members of the Torrence Park Citizens' Committee volunteered to conduct interviews in obtaining a random sample drawn to represent the total Torrence Park neighborhood. The survey sample consisted of 148 family interviews covering 671 persons, or a 30 to 31 percent sample of a total of 483 households and 2,100 persons. Data were earmarked for use by citizen groups in local and regional health planning activities, and to obtain appropriate and pertinent federal support where available in developing health-care services.

Survey data indicated widespread deficiencies in health-care services in the Torrence Park area, particularly for large families making less than $5,000 per year. Dental problems were quite evident, with over half of all families having had toothaches during the year, yet not seeing a dentist. Sixty to 70 percent seemed to obtain virtually no preventive dentistry.

Medically, about one-third had no family doctor; almost half of the low-income families neither saw a doctor for illness nor received a checkup during the year. Almost one-third of the children in these families saw no doctor during the year. This appeared generally consistent with other findings on the health needs of the poor, who generally tend to lose more time and suffer more illness than higher-income groups.

A major complaint of low-income groups and the elderly was transportation to the doctor's office. Many were also unable to obtain medical service. Those going out of town for medical care appeared to be more generally in the higher-income group, whereas the poorer ones seemed often not to receive any medical attention. The poorer groups also seemed often not able to obtain needed medicine. Similarly, almost two-thirds of the low-income families experiencing an emergency or accident appeared to have difficulty obtaining medical attention. A large percentage (66 percent) also would not seek medical help when they felt they needed it because of the expense. A considerable number of elderly shut-in persons were also represented in the sample.

Mental health services seemed predominately lacking, with as many as 80 percent of all groups never having used the mental health clinic. More than half of most groups did not know where to go for such help.

Nutritional advice was lacking throughout, particularly for the isolated aged. Over half did not know where to go for emergency rations.

Care of vision also appeared to be deficient, particularly for the lowest-income families. Almost half the children in the low-income group had had no preschool eye examination, with one in five of these not receiving such an examination even after entering school.

Almost half the families indicated no interest in birth control, while 80 to 100 percent indicated an interest in instruction on health.

Rodents seemed to be a common problem throughout the area,

with low-income groups indicating some problem in garbage pickup.

The Torrence Park survey subsequently served as the basis for forming a small, minimally funded (through Office of Economic Opportunity funds) health improvement operation for the neighborhood. It consisted of a small rental facility, VISTA workers, and several volunteer doctors. In further organizing and seeking additional funding sources, the operation has continued to expand. Guidance in management control operations was provided from technical and computer facilities available in the community which were offered gratuitously in response to the survey data highlighting the health needs of the poor.[16]

MENTAL HEALTH AND WELFARE
HUMAN-SERVICE SYSTEMS

Definitions often vary as to what constitutes a system; however, all seem to carry the common thread of an organized set of activities directed toward the solution of given problems within given environmental constraints. These are most easily illustrated in functional flow diagrams. (See Figures 7, 9, and 15 for examples of functional flow diagrams in the present text.)

Unfortunately, delivery of mental health and welfare services is virtually never presented in a functional flow sequence designed to achieve a desired outcome. Rather, when diagrams or overviews of an organization are prepared, these are developed in the form of what might be called "status" charts, or indeterminate systems in which physical parts do not correspond to separate functions designed to achieve given outcomes.[17] The federal government, for example, presents a set of status organizational charts, none of which demonstrate functional implications or activities that might be directed toward achieving given results (see Figure 32). This is likewise emulated by State government (see Figure 33 for example—a chart of the State of Maryland agency hierarchy). Indeed, when governmental agencies continue as separate unrelated entities, achievement goals are obscured and possible system developments may be exceedingly limited. Multiplying of agency entities in mental health and welfare suggests the building of inordinate‛ complexities and administrative barriers to the accomplishment of effective service delivery. Table 15 illustrates the number of separate agency entities across the states, often introducing administrative obstacles to effective functional flow of comprehensive services.[18]

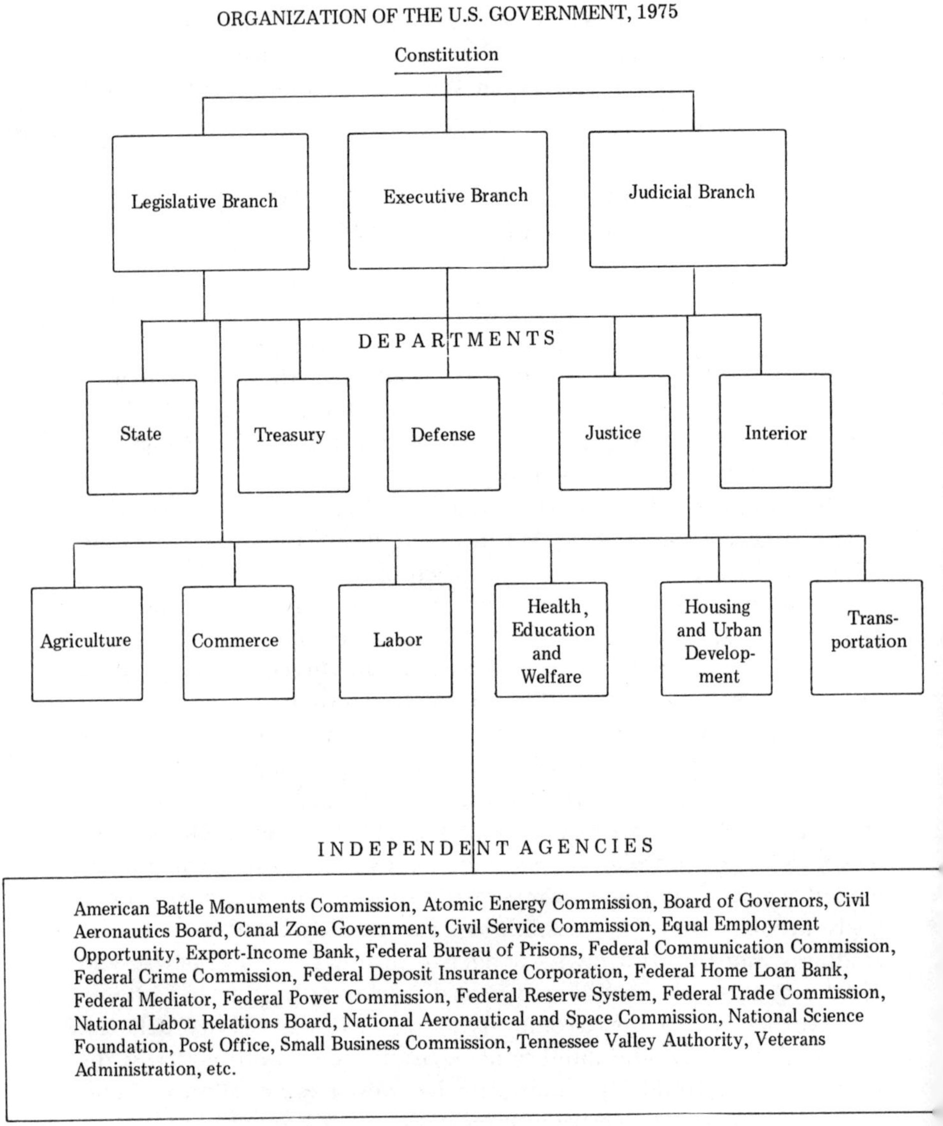

ORGANIZATION OF THE U.S. GOVERNMENT, 1975

Constitution

| Legislative Branch | Executive Branch | Judicial Branch |

DEPARTMENTS

| State | Treasury | Defense | Justice | Interior |

| Agriculture | Commerce | Labor | Health, Education and Welfare | Housing and Urban Development | Transportation |

INDEPENDENT AGENCIES

American Battle Monuments Commission, Atomic Energy Commission, Board of Governors, Civil Aeronautics Board, Canal Zone Government, Civil Service Commission, Equal Employment Opportunity, Export-Income Bank, Federal Bureau of Prisons, Federal Communication Commission, Federal Crime Commission, Federal Deposit Insurance Corporation, Federal Home Loan Bank, Federal Mediator, Federal Power Commission, Federal Reserve System, Federal Trade Commission, National Labor Relations Board, National Aeronautical and Space Commission, National Science Foundation, Post Office, Small Business Commission, Tennessee Valley Authority, Veterans Administration, etc.

Fig. 32. Illustrative Status Chart of the U.S. Government
(from U.S. Government Manual, 1974–1975)

ORGANIZATION OF THE GOVERNMENT OF THE STATE OF MARYLAND

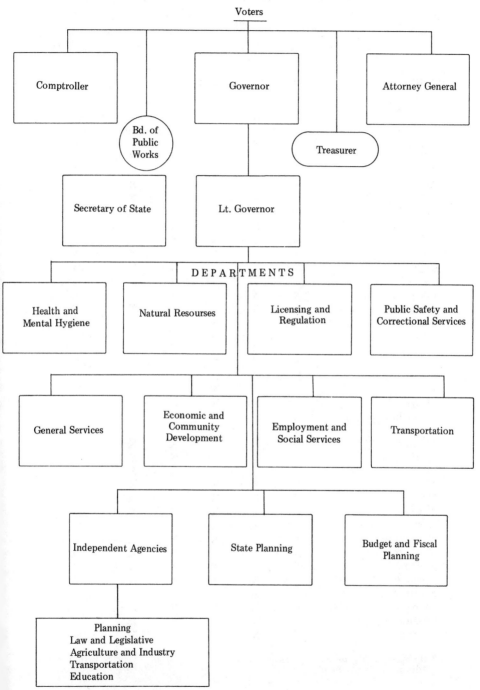

Fig. 33. Status Chart of the Government of the State of Maryland
(from Maryland Manual, 1971–1972)

TABLE 15

HUMAN-SERVICE CONGLOMERATES FOR HEALTH AND WELFARE SERVICES
(ADMINISTRATIVE UNIFICATION)

	Child. & Fam. Service	Mental Health	Wel- fare	Health	Voc. Rehab.	Correc- tions	Human-Service Admin. Score (HSAS)*
Alabama		X	X	X	X	X	.20
Alaska	X			X	X	X	.25
Arizona		X	X	X	X	X	.20
Arkansas		X	X	X	X	X	.20
California		X	X	X	X	X	.25
Colorado		X	X	X	X	X	.20
Connecticut		X	X	X	X	X	.20
Delaware		X		X	X	X	.25
Florida		X		X	X	X	.25
Georgia		X	X		X	X	.25
Hawaii				X	X		.50
Idaho		X	X	X	X	X	.20
Illinois	X	X	X	X	X	X	.16
Indiana		X	X	X	X	X	.25
Iowa		X	X	X	X	X	.25
Kansas		X	X	X	X	X	.25
Kentucky	X	X	X	X	X	X	.16
Louisiana		X	X	X	X	X	.20
Maine		X		X	X	X	.25
Maryland		X	X	X	X		.25
Massachusetts		X	X	X	X	X	.20
Michigan		X	X	X	X	X	.20
Minnesota		X	X	X	X	X	.20
Mississippi		X	X	X	X	X	.20
Missouri		X	X	X	X	X	.20
Montana		X	X	X	X	X	.20
Nebraska		X	X	X	X	X	.20
Nevada		X	X	X	X	X	.20
New Hampshire		X	X	X	X	X	.20
New Jersey		X	X	X	X	X	.20
New Mexico		X	X	X	X	X	.20
New York		X	X	X	X	X	.20
North Carolina		X	X	X	X	X	.20
North Dakota		X	X	X	X	X	.20
Ohio		X	X	X	X	X	.20
Oklahoma		X	X	X	X	X	.20
Oregon		X	X	X	X	X	.20

* HSAS = Unity of Administrative Control, or 1/ (No. of Agencies with Separate Administrative Control)

TABLE 15 (CONT'D.)

Pennsylvania		X	X	X	X	X	.20
Rhode Island		X	X	X	X	X	.20
South Carolina		X	X	X	X	X	.20
South Dakota		X	X	X	X	X	.20
Tennessee		X	X	X	X	X	.20
Texas		X	X	X	X	X	.20
Utah	X	X		X	X	X	.20
Vermont		X	X	X	X	X	.20
Virginia		X	X	X	X		.20
Washington				X	X		.50
West Virginia		X	X	X	X	X	.20
Wisconsin		X		X	X	X	.25
Wyoming		X	X	X	X	X	.20
District of Columbia		X	X	X	X	X	.20
American Samoa		X	X	X	X	X	.20
TTPI		X	X	X	X	X	.20
Virgin Islands		X	X	X	X	X	.20

Based on number of separate agencies providing discrete service by category, a "Human Service Administrative Score" (HSAS) was devised for mental health and welfare services. Several states, for example, have disaggregated agencies for child abuse and neglect (which is a highly restricted charter). Such a practice may be taken to downgrade the total human-service administration of comprehensive mental health and welfare services, thus reducing the total score. A perfect score, where all services fall under one administrative body, would be 1.00. Scores in this analysis range from .16, where all welfare services are disaggregated administratively, to .50, where only two administrative bodies are indicated for mental health, physical and social welfare.[19]

CENTRALIZED CARE-GIVING DEVELOPMENTS

Widespread interest in human services is becoming evident, following the commonly recognized inadequacies of existing categorical, charter-restricting services. Indeed, a number of local organizations, sometimes with university affiliation, with variable scope of operation, have sprung up across the country.[20] Federal interest, though spasmodic and lacking extensive executive and legislative support, has also been evident as in the proposed Allied Services Act early in the seventies. A number of states have also indicated an interest in organizing human services:

Alaska: The Anchorage Borough Health Department has taken steps in planning human service delivery systems.[21]

Arkansas: A pilot project in rural Arkansas was employed to derive recommendation on structured management training, and information systems on participating organizations and consumer attitudes.[22]

California: Several counties or regions are combining administrative functions. San Mateo County, for example, is proposing a combined social services and welfare, health and mental health administration. Similar reorganization at the state level is also being contemplated.[23]

Delaware: The State of Delaware is proposing a series of service centers across the state to include some measure of the following agency involvement: legal, social, health, mental health, retardation, corrections, vocational rehabilitation, employment, drug abuse, youth opportunity, planned parenthood, courts, human relations, and consumer affairs.[24]

Florida: A comprehensive pilot service delivery system was begun in Florida in 1969. Centralization and computerization problems have continued to be formidable, however.[25]

Georgia: The Department of Human Resources was instituted in 1973 to combine such separate agency functions as health, children and youth vocational rehabilitation, and children and family services.[26]

Kentucky: The Kentucky Public Health Association recently concluded that medical care will continue to be fragmented and ineffective until combined with social services. The association recently made recommendations to the state legislature to develop incentives for health-care specialists to work in poverty areas, and to encourage members of poverty groups to enter health, etc., fields.

Michigan: A tri-county project is being employed to pilot and test intake and case management processes.[27]

Massachusetts: The State Executive Office of Human Services underwent a proposed reorganization by the Governor in 1973. However, administrative consolidation did not appear to be a major design objective.

Pennsylvania: A consortium of agencies for child care and protection in the Pittsburgh area is being developed under federal grant. Child guidance, welfare, juvenile court and children's homes are involved, ultimately to provide broader family-based services.

Washington: The Department of Social and Health Services is looking at a total human-services delivery system, in the interest of expanding comprehensive community mental health services at the local level.[28]

Wisconsin: Planning is underway in the Division of Mental Health for development of integrated and coordinated services to the disabled in this category—to include a unified information system for use in planning.[29]

The mental health and welfare mess, as it is often aptly alluded to, was dramatically described by the columnist Carl Rowan:

. . . The simple reality is that poor, jobless, ill-fed people get sick a lot, which is why Medicaid payments jumped 22.3 percent to $13 billion. If . . . policies [are] pursued that leave people out of work, force them to have babies they don't want and can't afford, and consign them to certain ill health [there will be] one whopping welfare bill. . . .

. . . school officials will push a million kids into the streets . . . those who are a little too active for teachers who like placidity; kids who are a little extra slow at grasping the new math; kids who pop bubble gum . . . and make passes at 12-year-olds of the opposite sex, who show up at school with knives, brass knuckles, even guns. . . .

. . . Who needs an extra-special effort by compassionate school officials more than these troubled kids? . . . Those kids are doomed to populate this country's 3,921 local jails; to move onward into our already bulging prisons, to become the rapers and robbers, the faceless people who lie behind the crime and welfare statistics that will outrage us 10 to 15 years from now . . . while no real incentive is offered to leave welfare and take a modest job. . . . One welfare mother who opted for hard work wound up making exactly $2 more per month than welfare paid. . . .

METHODS IN THE ANALYSES OF FRAGMENTED SYSTEMS

Often various facets of the commercial, domiciliary, criminal justice, welfare, etc., subsystems may be isolated for study of problem areas in operations as described in the foregoing examples. Study approaches may concern a simple examination of such separate and isolated problem areas as described, or a subsystem may be subjected to more total systems study. In order to accomplish a more comprehensive approach to the development and implementation of system objectives, the following system study procedures may be employed as discrete steps in troubleshooting current operational systems:[30]

1. Identify goals over which each specified concerned individual has administrative jurisdiction.
2. Prepare block diagrams of current operations that contain the

components that purport to accomplish the goals, or that obscure the operational goals.

3. Identify key actors/performers in each operational role.
4. Prepare structured interviews for substantive penetration into each component of the subsystem.
5. Identify each trouble area where system flow or effective performance is blocked.
6. Prepare a problems summary and recommendations.
7. Coordinate recommendations through appropriate administrative personnel under whose jurisdiction each of the outcome goals fall.

While human-service systems for the foreseeable future may be uncoordinate, lacking controlled total systems development direction that is aimed at immediate and long-range alleviation of misery, inefficiencies. and excessive costs, recognition of these deficiencies becomes increasingly widespread. The glowing awareness of the separated, costly and counterproductive facets of human-service design may well lead to new and aggressive systems design approaches. Federal leadership is needed above all else, that is both compassionate and "hard-nosed," that is sensitive to, yet realistic about, the turmoil and distress of our cities and the plight of our citizens—a leadership that will guide and point the way with necessary technology and through proper and effective incentives to states, institutions, and citizens.

Many of the methods are currently available. These may be explored and identified, and, through new as well as proven methods of analysis, we may seize upon opportunities to advance technology in the most critical areas of human concern.

NOTES TO CHAPTER 13

1. A general hospital psychiatric placement project in Illinois required such cooperation. The Department of Mental Health administered the project, while Public Aid monies were used to finance it. Only marginal success was possible, however, due to the extended delays incurred in payments to hospitals and private physicians.

2. A population of rural southern origin in their native habitat had habitually heaved garbage out the window to feed the hogs. When participating in a northern urban housing project, this habitual mode of behavior created sanitation and living problems. Garbage disposal and training would thus require special attention for this population.

3. Noise pollution is basically a multiple series of vibrations through media in a frequency range to which the human ear and body, as well as building structures, are responsive. The ear responds from about 20 through 20,000 hz or cycles per second. The ear is most sensitive in the range from 2,000 to 4,000 hz,

where 0.001 dynes per square centimeter will evoke a response. At the extremes of the audible range, pressure changes 3,000 times as great are required. The decibel (db) (a term borrowed from electrical communication engineering) is the relative quantity measure used to describe magnitude or variations in pressure, e.g., at a given distance, 30 db may be measured from a soft whisper as against 125 db for a jet aircraft taking off. Based on the differences in perceived loudness (measured in "phons") as a function of hertz, the American National Standards Institute specification for Sound Level Meters requires that instruments provide three frequency response characteristics designated A, B, and C. The A-weighted measure (dbA) most nearly corresponds to hearing. For more detailed discussion of noise characteristics, refer to A. Peterson and E. Gross, *Handbook of Noise Measurement* (Concord, Mass.: General Radio Company, 1973).

4. T. Schultz, "Noise Assessment Guidelines Technical Background," *U.S. Department of Housing and Urban Development* HUD Report No. TE/NA 172 (Washington, D.C., 20410, 1971).

5. See, e.g. W. Connor et al., "Community Reaction to Aircraft Noise Around Smaller Airports," *National Aeronautical and Space Administration*, CR 2104 (Washington, D.C., 1972).

6. Complaints are active, possibly organized, reactions with quasi-demands that something be done about the noise. Annoyances are usually uncovered in surveys questioning representative samples of exposed populations. Complaint frequency occurs in a ratio roughly proportional to the square root of annoyance frequency.

7. *Criminal Justice Newsletter* (Continental Plaza, 411 Hackensack Avenue, Hackensack, N.J. 07601: National Council on Crime and Delinquency, November 18, 1975).

8. D. Patrick et al., "Toward an Operational Definition of Health," (Health Index Project Department of Community Medicine University of California, San Diego/La Jolla, California, 1972).

9. The Illinois Health Department, for example, compiles weekly and annual statistics on such diseases as German measles, hepatitis infection, tuberculosis, syphilis, etc. Though many such, of course, go unreported, these official reports offer some basis for planning preventive programs.

10. R. Aldrich and R. Wedgwood, "Health Services for Children and Youth," (Health Resources Study Center, Department of Pediatrics University of Washington, School of Medicine, Seattle, Washington, 1970).

11. J. Stanfield, "Some Human Factors Aspects of Emergency Medical Care" (Proceedings of the Seventeenth Annual Meeting of the Human Factors Society, The Human Factors Society, P.O. Box 1369, Santa Monica, California, 1973) pp. 200–205.

12. R. Zitnik (Director of Cardiology), Computer-Processed ECG, Little Company of Mary Hospital, Evergreen Park, Chicago, Illinois, 1974.

13. S. Lippert, "Travel in Nursing Units," *Human Factors Journal* 13 (June 1971) :269–82; P. Ronco, "Human Factors Applied to Hospital Patient Care," *Human Factors Journal* 14 (October 1972) :461–70.

14. Department of Health, Education and Welfare, Office of the Assistant Secretary, Planning and Evaluation, "Human Investment Programs, Delivery of Health Services for the Poor," (Washington D.C.: U. S. Government Printing Office, 1967).

15. "Era of the HMO: Health Care by the Year," *Science News* 105 (January 19, 1974) :38 f.

16. P. Tobias et al., "Human Factors in the Design of a Computerized System for a Neighborhood Health Clinic," (Proceedings of the Sixteenth Annual Meeting of the Human Factors Society, The Human Factors Society, P. O. Box 1369, Santa Monica, California, 1973) pp. 247 f.; R. Herzlinger et al., "Management Control Systems in Health Care," *Medical Care* 9 (September 1973) :416–29.

17. J. Jones, "The Design of Man-Machine Systems," in *The Human Operator in Complex Systems*, ed. W. Singleton et al. (London: Taylor & Francis Limited, 1967) , pp. 1–11.

18. Council of State Governments, State Administrative Officials Classified by Functions. The Book of the States (Iron Works Pike, Lexington, Kentucky 40505, 1973) .

19. Vocational Rehabilitation, which is primarily a federally sponsored agency, has been assumed to be functionally disaggregated administratively across all states. This is often grouped administratively under the Department of Public Instruction. Here, too, the charter is highly restricted administratively, serving only the physically or mentally disabled over 16 years of age who are judged able to benefit from vocational training.

20. Refer, for example, to the National Association of Human Service Technology, 1127 Eleventh Street, Sacramento, California 95814; The Center for Human Services Research, 212 Harriet Lane, Baltimore, Maryland 21205; The Academy of Human Service Sciences, 30 North Brainard, Naperville, Illinois 60540; The Human Service Department, Brandeis University, Waltham, Massachusetts; The Human Services Research Laboratory, Case Western Reserve University, Cleveland, Ohio 44106.

21. Personal correspondence with J. McClain, Department of Health and Social Services, Pouch 14, Juneau, Alaska 99801, 1973.

22. Arkansas School of Social Work, "An Evaluation to Determine the Effectiveness of Coordinated Administration and Delivery of Service by a Multi-Service Center in Rural Arkansas" (Arkansas University, Fayetteville, Ark., 1973) .

23. Nancy Crabbe, Department of Public Health and Welfare, Personal correspondence with Librarian, 225 37th Avenue, San Mateo, California 94403, July 1973.

24. Charles Debnam, Division of State Service Centers, Personal correspondence, New Castle, Delaware 19720, 1973.

25. Florida State Department of Health and Rehabilitation Services, "Comprehensive Services Delivery System. Its Nature and History" (Tallahassee, Florida, 1974) .

26. Georgia Department of Human Resources, 47 Trinity Avenue, Atlanta, Georgia 30334.

27. Michigan Department of Management and Budgets, "Common Intake Case Management Pilot Implementation: Management/Administrative Manual" (Governor's Human Serivce Council, Lansing, Mich., 1974) .

28. W. E. Lehman, Washington State Office of Mental Health, Department of Social and Health Services, Social Service Division, Olympia, Washington 98504, 1973.

29. K. Hyre, personal correspondence, Planning Division of Mental Hygiene (Department of Health and Social Service, Madison, Wis. 53702, 1973) .

30. Refer to examples described in the study of BARTD in chapter 11, and geriatric placement in chapter 4.

14

Human-Service Systems Integration
and Evaluation

Systems integration, of course, is a broad designation encompassing all system components and elements. In the need for integration, it is essentially maintained that all elements must perform within required maximal outside limits of tolerance for total system outcomes to be accomplished.

Systems integration, as a required design process, properly is a systems engineering term. A human-factors analysis must also be regarded as an essential part of a total systems analysis, which must begin with systems goal specifications, and end with the maintaining or iterating of required outcome measures. A simplex model of systems integration and evaluation is presented in Figure 34.

DISCLOSURE OF ISSUES

The systems or human-factors analyst might often be most behooved to search out, clarify and define issues and goals of human-service projects. Indeed, in the modern management of industry and commerce, it becomes quite essential to adopt goal-orienting practices to enhance motivation of staff and organize activities about desired outcomes.[1] Theoretical goal-directed approaches to service are, in fact, becoming increasingly popular among human-service agencies and in state government.[2] In practice, however, the outlook for carefully planned and integrated human-service projects is often seen as discouraging.[3] The planning and evaluation phase in state government administration and other levels of human-service management is frequently and paradoxically decimated in budgetary crunches at a time when they are precisely most needed to establish fiscal priorities.[4] King[5] laments that in our age of analysis when personality integration

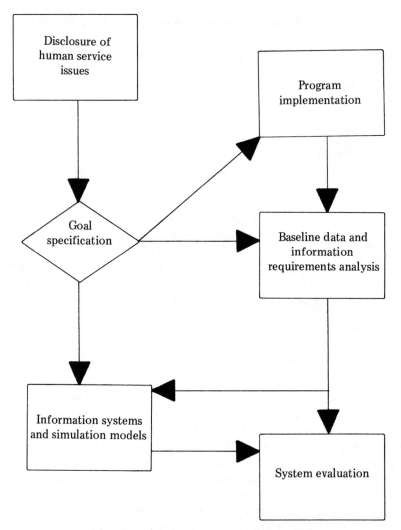

Fig. 34. A Simplex Model of a Human-Service Systems
Integration-and-Evaluation Process

becomes a major objective, we often most neglect integration of the social milieu, from which personalities derive, in psychiatric services. It can, in fact, be seen that integration of most human-service projects is often the most neglected aspect of project design.

The lack of definitive goal-oriented human-service projects and the concomitant lack of initiative by government and agency officials to establish such, perhaps presents the major barrier to the effective employment of system approaches to the human-service industries. A number of methods, however, might be employed by systems and human-factors engineers to assist in the clarification of development issues in establishing meaningful and valid goals. These include surveys on public attitudes and annoyances, e.g., on the beautification of public facilities, rapid inner city transit, slum clearance, etc. Intensive interviews of public officials and agency managers (e.g., employing personal in-depth interview techniques, or the DELPHI technique as developed by the Rand Corporation, etc.) may also be accomplished to obtain consensus data from which to derive a disclosure of issues.

A recent method developed for mental-health volunteer coordinators, but applicable to a more general population of managers, was that of a management-by-objectives workshop. Participants were asked to described specific projects, populations to be served, and general outcome goals for the project. From there, detailed objectives were worked out relative to legitimate baseline data. General programming plans were identified, as were the constraints and other administrators or public officials involved in the project who must be convinced of its importance. For example, one volunteer coordinator specified a 50 percent increase in participation of high school volunteers in chronic schizophrenic wards. Baseline data consisted of the number of such volunteers participating during a previous period. Constraints were identified as the limited released-time available from normal school activities and the required cooperation of parents and school administrators. Others involved in the program consisted of ward personnel under whose direction the students were required to work.[6] For assessing progress in achieving objectives, periodic data review was specified, as well as the identity of those participating in the review for whom the goals and review data were to be meaningful. The techniques for review were also specified in terms of oral and written reports, graphs, charts, etc.

The general worksheet employed in the workshop is presented in Exhibit B. Such practicum workshops might be employed extensively for orienting managerial personnel to meaningful project design goals, their integration and outcome evaluation.

EXHIBIT B

MANAGEMENT-BY-OBJECTIVES (MBO) WORKSHEET

The following exercise is designed as a brief practicum for an MBO application based on management problems in your own organization. Please complete each area for discussion in as much detail as possible. Use additional sheets as necessary.

1. Describe Service Project:
2. Target Population:
3. General Goal Statement:
4. Specific Objective (s) Within Project:
5. Describe Baseline Data Required:
6. Necessary Programs or Activities to Accomplish (4) Above:
7. Constraints or Obstacles to be Overcome:
8. Others Who Must Participate or Become Involved to Accomplish Objectives:
9. Review Periods for Monitoring or Feedback:
10. Persons Involved in Review:
11. Techniques for Presentation and Review:

MANAGEMENT TOOLS IN HUMAN-SERVICES SYSTEMS INTEGRATION

For nearly a decade the systems approach has been espoused for the design, management, and evaluation of human-service operations. In the early sixties, the then Defense Secretary, Robert McNamara, acclaimed the systems methods of analysis over those of management judgments based on intuition or experience. The better the factual basis for reflective and projective judgment, the better the overall judgment would be, insisted McNamara.[7] However, during the intervening years of such methodological espousal, but few managers and purveyors of human-services acquired the essential skills to implement such approaches. The result has largely been continued mismanagement and fragmentation in human-service systems. In contemporary tradition, for example, psychiatric expertise is assumed to be the appropriate background for mental health administration, viz., a physician-administrator to direct patient management and policy formulation concerning treatment and incarceration practices.[8] Psychiatric administrative functions have come to establish medical definitions of virtually all mental health problems, together with the nature of services and modes of delivery to be provided, including types of activities and relationships, staff roles and status in regard to inpatients or patients in the community. When administrative skills do become an issue in the psychiatrist's training curriculum, it is addressed more as a kind of clinical problem analogous to a macrocosm of illness.[9]

Through the force of such extraneous factors as medical prestige and generalized vested interest of the medical peer group, the psychiatrist, though normally an inept adminstrator, is thus maintained in his role to exercise a mandate administratively with only randomly effective administrative skills to manage the institution and community psychiatry programs. As an administrator, he thus imposes the limiting effectiveness of his total medical perspective, even while current controversies in mental health center about the appropriateness of the medical model in treatment when human needs are the more cogent issues bearing on community acceptance.[10]

Integration of such human services may become more basically an administrative than clinical problem for innovation in addressing essential human needs. The administrative skills needed to embark on such innovation may be seen to include the following administrative functions:

Personnel Management: To select appropriate personnel, and maintain a level of motivation commensurate with the working consistently toward goals and subgoals of the pilot project.

Planning: Developing essential specifications and mission objectives of the human-service pilot project through an assemblage of appropriate teams of specialists. Scheduling time for implementation, testing and evaluation of results, cost estimating to completion through Gantt charts, etc.

Computer Use and Justification: Performing preliminary trade-off studies to establish that volume and access to invisible data through modeling and simulation will enhance planning and evalution procedures, and result in cost saving.

Documentation: Maintaining records and mission-oriented research data, and project progress to determine problem areas and solutions.

Performance Evaluation Review Techniques and Critical Path Method (PERT/CPM): Maintaining close monitoring surveillance over project progress, and critical paths to completion, periodically and regularly rescheduling manpower and resources to maintain schedules to project completion.

Operations Research Studies and Review: Directing and reviewing findings to establish costs and benefits to be derived from the innovative studies in the pilot project, through the employment of resources on a broader operational scale and the evaluation of outcomes, e.g., a dispersion of advocates and service providers in the community versus an institutional setting, etc.

Such administrative skills would, of course, be highly desirable in human services. Perhaps rather more typical of human-services administration today, however, is the propensity to focus on more limited administrative processes, rather than on innovative programming and integration about broad outcome goals.[11] Indeed, modern human-service managers tend to turn to computer technology, often indiscriminately to promote managerial prestige, if not functional program management. Van Cott and Kinkade[12] have attempted to establish more functional user needs for computers by simulating first the computer functions before committing such resources to data bases of doubtful utility. Other studies have dealt with efficiency of interaction with computers and the accessibility and response time of time-shared systems.[13]

Peace and Easterby[14] have attributed the widespread use and misuse of computers to a paucity of human-factors systems input in the development and design of computer functions.

Weissman[15] maintains that ineffective management in human-service agencies may often require drastic measures in removing ineffective administrators to bring about needed reforms. Such ineffectual administration, however, appears still to be but a manifestation of a continuing lack of aggressive definitions, through public opinion and government executive initiative, of policies about which goals might be integrated. With such operational policies, broader-based outcome goals might be brought into clear focus, while the ineffective agency administrator and those with bureaucratic rigor mortis in general may then become converted to more functional resource personnel integrating their efforts about dynamic innovative human-service programs.

PROJECT INTEGRATION IN MENTAL HEALTH—AN ILLUSTRATIVE PROJECT

During 1973 a plan was developed for effective administration and integration of an aftercare project for mental patients in Champaign County, Illinois. Effective aftercare programs for mental patients discharged from mental hospitals or psychiatric wards of general hospitals must assure that appropriate, timely and effective follow-up services are delivered to prevent relapse and obviate the return of the former patients to an inpatient status. The administrative project essentially concerned the control of Department of Mental Health (DMH) allocated funds to various community agencies or human-service functions to assure the effective delivery of aftercare or "sustaining care" services.

The Champaign County catchment area in Illinois consisted of approximately 200,000 in population, having urban areas of approximately 160,000 in a surrounding of sparsely populated rural farming area. The public charge of the State Department of Mental Health was largely to provide mental health inpatient services and sustaining care for discharged mental patients within this catchment area, with primary emphasis on community-owned services. The administrative functions in managing or integrating state programs in the community concerned the controlling of state monies (up to 40 to 60 percent of total funding) to assure delivery of appropriate and effective services by community agencies.

Overall objectives were essentially to reduce required levels of patient care based on current baseline data. Table 16 outlines the various levels of care operating in the Champaign catchment area with base year patient volume at each level. Measurable objectives, about which state monies were negotiated with 13 distinct service agencies, concerned decreased volumes at the higher levels of care. For example, a patient group in state hospitals (Level V) were negotiated for placement in lesser-care night treatment services (Level IV) ; or outpatients requiring extensive coordination and advocacy (Level II) were negotiated for a reduced level of crisis episodic care (Level I). Each negotiated outcome, of course, required a carefully planned course of action, situational or treatment strategy.

Figure 35 outlines the negotiation and integration process in program administration. Initial Data Input (1) consisted of current volumes by level of care as presented in Table 16. Desired Outcomes (2, Figure 35) were negotiated with each agency, where necessary, by individual patient. Periodic review permitted iteration or Data Feedback (3) to assess progress and trends. The decision question, "Are Goals Being Met?" (4), provided the direction for conducting surveys to determine conditions of failure (5), continuing payment and/or renegotiating for improvements (5a). In negotiating for improved or corrected service functions, or new control strategies (5b), models were to be constructed for analysis. The network analysis model, for example, as described in chapter 6, provided an overview of patients passing through various service networks, and permitted measuring cycles or reverberations within the system that marked deficiencies

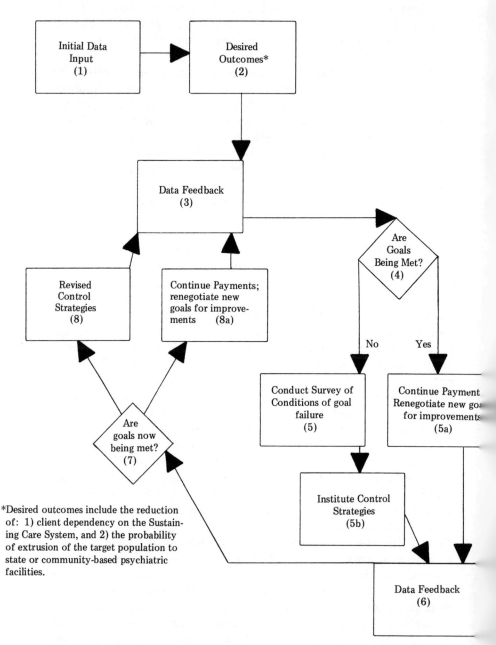

Fig. 35. Functional-Flow Diagram of an Information and Control System

TABLE 16

LEVELS OF CARE FOR PSYCHIATRIC PATIENTS IN
THE CHAMPAIGN CATCHMENT AREA, PER ANNUM

Level of Care		Current Volume (Baseline Data)
I. Episodic Care		3925
A. Outpatient Care	2215	
B. Crisis Intervention	1710	
II. Outpatient Case Coordination		430
III. Day Programs		179
A. Sheltered Workshop	95	
B. Day Treatment	72	
C. Work Activity	12	
IV. Night Treatment		270
A. Three-quarter Houses (supervised, independent living)	50	
B. Halfway Houses (supervised short-term residency)	95	
C. Community Group Living Facilities (supervised long-term residency)	125	
V. Twenty-Four-Hour Treatment		686
A. Short-Term Psychiatric Inpatients in General Hospital)	570	
B. Short-Term Inpatients in State Hospitals	100	
C. Long-Term Inpatients in State Hospitals	16	
Total Number of Patients per Annum		5490

or inadequacies of service. Such further Data Feedback (6) was employed to ask the further question if goals are now being met (7), with continuing revision of control strategies (8) and /or agency renegotiation (8a).

The project-integration cybernetic model, as described above, was designed specifically for iteration on services for post-hospitalized mental patients. However, in principle the model could likely be adapted to any of a number of human-service systems. Such models could relate to recidivism in the criminal-justice subsystem, reduced false-alarm frequency in fire protection, reduced congestion frequency in surface traffic, parts-per-million of contaminants in the sanitation subsystem, etc. The outlined procedures, of course, presuppose elements of control that can be manipulated by the project administrator; and, in the case of the more complex systems, it is assumed that scientific models can be constructed to describe, assess, and predict outcomes.

OUTCOME EVALUATION

The human-factors system or project development cycle, as described by Meister and Rabideau[16] involves a predesign or planning phase, detailed design, a production, fabrication or implementation phase, and a design-verification, evaluation and product-improvement phase. Human-service systems, as iterated throughout the course of the present text, may unfortunately not lend themselves easily to definitive verification, often lacking clearly specified mission objectives. In fact, in the absence of a knowledgeable and goal-seeking administration, evaluative measures continue to be elusively unattainable. Broad-aim programs clearly do not lend themselves to experimental evaluation, while the failure to specify human-service objectives effectively means that integration and systems discipline is not possible.[17]

Meanwhile, information systems may continue to develop at various levels of administration, often without regard to broadly-based policies, goals or issues. The result is a group of independent information or computer applications, virtually *non compos mentis* when addressing a systems problem; nor is there opportunity for systems integration.[18]

In many human-service areas, indeed, conscientious proponents seem constantly to be casting about for meaningful evaluation methods and criteria. These may range from asking mental health patients or welfare recipients themselves how they feel about the service they received, the development of complex formulas relating to income-earning capability before and after the service, or a goal-attainment scoring system predicated on arbitrarily established ad hoc mini-goals.[19] Unfortunately, all such evaluation efforts appear to fall wide of the mark, when they seem more to be seeking justification for services being delivered rather than to update, improve, correct or rework ineffective service systems. Often feverish, though innocuous, evaluation efforts become meaningless exercises unless definitively circumscribed by social policy, the explication of goals, and a clear-cut specification of solutions being tested.

NOTES TO CHAPTER 14

1. P. Drucker, *The Practice of Management* (New York: Harper and Row, Publishers, 1954) ; R. Mockler, "Theory and Practice of Planning," *Harvard Business Review*, March 1970, pp. 148–58; L. Peter, *The Peter Prescription* (New York: William Morrow and Company, Inc., 1972) ; P. Crosby, *The Art of Getting Your Own Sweet Way* (New York: McGraw-Hill Company, 1972) .

2. H. Gottesfeld et al., "Strategies in Innovative Human Service Programs" (New York: Behavioral Publications, 1973) ; J. Haldane, "Human Factors and the Environment," *Proceedings of the Seventeenth Annual Meeting of the Human Factors Society* (The Human Factors Society, P. O. Box 1369, Santa Monica, California 1973) , pp. 292–98; Illinois State Department of Children and Family Services, "Directions '74," vol. 1, September 1973.

3. "About Psychiatry, Psychiatrists and Social Problems," *American Journal of Psychiatry* 130 (February 1973) :204–5.

4. H. Demone and D. Harshbarger, "The Planning and Administration of Human Services" (New York: Behavioral Publications, 1973) .

5. C. King, "On Making Relevance Relevant," *American Journal of Orthopsychiatry* 43 (October 1973) :717–19.

6. K. Greider and J. Burgess, "Released-Time High School Volunteers in a Chronic Schizophrenic Program" (Adolf Meyer Center, Decatur, Ill., 1974) .

7. "Systems Approach: Political Interest Rises," *Science* 453 (September 9, 1966): 1222ff.; R. Elkin, "Systems Approach to Managing Welfare Programs," *Social Work Practice*, 1968, pp. 159–74.

8. K. Menninger, "The Approach to the Psychiatric Patient," in *A Psychiatrist's World* (New York: The Viking Press, 1959) , pp. 378ff.; M. Foucault, *Madness and Civilization: A History of Insanity in the Age of Reason* (New York: Pantheon Books, 1965) .

9. M. Greenblatt, "Administrative Psychiatry" *The American Journal of Psychiatry* 129, no. 4 (1972) :373–86; E. Pattison, "Residency Training, Issues in Community Psychiatry," *The American Journal of Psychiatry* 128, no. 9 (1972) :1072–02; R. Becker et al., "Psychiatry in the Functionally Organized Undergraduate Curriculum," *The American Journal of Psychiatry* 30, no. 5 (1973) :571–73.

10. N. Farberow, "The Crisis is Chronic," *American Psychologist* 28, no. 5 (1973): 388–94.

11. M. Chesler and N. Flanders, "Resistance to Research and Research Utilization: The Death and Life of a Feedback Attempt," *Journal of Applied Behavior Science* 3, no. 4 (1967) :469–87; L. Kurke and H. Van Houdnos, "Staff Activity Reporting from the Bottom Up," in *Progress in Mental Health Information Systems: Computer Applications*, ed. J. Crawford and D. Morgan (Cambridge, Mass.: Ballinger Publishing Company, 1974) .

12. H. Van Cott and R. Kinkade, "Human Simulation Applied to the Functional Design of Information Systems," *Human Factors Journal* 10 (June 1968): 211–16.

13. C. Morrill, "Computer-Aided Instruction as Part of a Management Information System," *Human Factors Journal* 9 (June 1967) :251–56; R. Root and R. Sadacca, "Man-Computer Communication Techniques: Two Experiments," *Human Factors Journal* 9 (December 1967) :521–28; R. Nickerson et al., "Human Factors

and the Design of Time Sharing Computer Systems," *Human Factors Journal* 10 (April 1968) :127–34.

14. D. Peace and R. Easterby, "The Evaluation of User Interaction with Computer-Based Management Information Systems," *Human Factors Journal* 15 (April 1973) :163–77.

15. H. Weissman, *Overcoming Mismanagement in the Human Service Professions* (San Francisco, Calif.: Jossey-Bass, Inc., Publishers, 1973) .

16. D. Meister and G. Rabideau, *Human Factors Evalution in System Development,* (New York: John Wiley and Sons, Inc., 1965) .

17. R. Weiss and M. Rein, "The Evaluation of Broad-Aim Programs: Experimental Design, Its Difficulties, and an Alternative," *Administrative Science Quarterly* 15 (1970) :97–109.

18. R. Van Dussedorp, "Some Principles for the Development of Management Information Systems," in *Management Information Systems for Higher Education: The State of the Art,* ed. C. Johnson and W. Katzenmeyer (Durham, N.C.: Duke University Press, 1969) .

19. J. Burgess, "Mental Health Service Systems: Approaches to Evaluation," *American Journal of Community Psychology* 2 (1974) :87–93.

15

The Future of Systems Approaches
in Public Services

... I have nothing but contempt for the kind of governor who is
afraid, for whatever reason, to follow the course that he knows is
best for the State; and as for the man who sets private friendship
above the public welfare—I have no use for him either ..." Sophocles,
495–405 B.C.

... Until ... political greatness and wisdom meet in one, and those
commoner natures who pursue either to the exclusion of the other
are compelled to stand aside, cities will never rest from their
evils ..." Plato, 428–348 B.C.

... The Commons, faithful to their system, remained in a wise and
masterly inactivity ..." Sir James Mackintosh, 1765–1832.

In the decade of the 1970s we have been subjected to a new lexicon
of crises—the energy crisis, the food crisis, the moral crisis, the population crisis, the resource crisis, the rural crisis, the urban crisis, etc.

Social prophets frequently hold the grim argument that the future
can be seen only as a continuation of the darkness, cruelty, insensi-
tivity and disorder of the past. Threats to survival are seen in runaway
population growth, obliterative war, and the exhaustion of the environment. Famine and disease may continue to be the major source
of population control in the underdeveloped countries, while the
threat of major war between the big powers may give way to a nuclear
disbalance of power when nuclear weapons are wielded by small
nations as a means to redistribute the wealth. Industrial heat and
other man-made thermal output at present growth rates in industry
may soon create a dangerous margin in heat rise, producing cata-
strophic climatic change within a few generations. We may indeed
face multiple convulsive changes forced upon us through the breakdown of such environmental and political system balances.

The Department of the Interior has calculated that the United States, with 8 percent of the world's population, uses half of the world's material resources. Copper, sulfur, and phosphates are the only materials adequately produced internally; whereas iron, aluminum, nickel, lead, zinc, tin, chrome, tungsten, etc., must be imported for 30–100 percent of the nation's requirements. The smaller nations owning such resources may strategically control their distribution for inordinate profit and political advantage. Meanwhile, for all but coal and iron, overall exhaustion of resources throughout the world, as with oil, may occur within a century.

Pressing urban needs and congestion also continue as increasing problems, with over 30 million Americans moving to urban and suburban areas since 1940. Such crowding confounds and multiplies the need for human services and/or major and radical solutions to relieve the seemingly insurmountable city problems of the future.

The reality and scope of these problems, moreover, is only partially understood or verifiable for, as Toffler[1] has noted, the only comprehensive social indicator conscientiously developed for monitoring is the Gross National Product. We indeed have no food consumption index, nor resource deficiency, nor urban blight, environmental census, etc., indices to provide measures as to whether or not the country is more or less livable from one year to the next. A major factor in resource consumption rates, for example, is the number of consumers. By gross calculations there will be twice as many as the three and a half billion people on earth today before the turn of the century. We can only project, on the basis of current trends, that within a matter of decades massive famines will be upon us.

Solutions must lie in careful study and projections. In resources, for example, the National Commission on Materials Policy postulated that a global crisis will soon develop if there is not a movement toward international cooperation in the exploitation, development, control and distribution of world resources. Urban crises, according to Dr. Peter Goldmark,[2] may be averted through the freeing of people from the economic necessity of living near large cities. This, he has studiously asserted, may be accomplished by allowing them to move to small country towns linked electronically to entertainment, business, medical, government, etc., centers. The technology, he contends, is already available and requires only imaginative applications. Many observers in our time see future survival, in the face of overwhelming human-service problems, to be possible only under governments capable of rallying obedience far more effectively than is possible in a democratic setting. When the National Science Foundation was established in 1950, its charter gave the physical sciences the major emphasis, while neglecting, until recently, urgent social needs. Today the question is still unanswered if our current social structures and political systems are compatible with use of the scientific method for solving complex human-service problems.[3] The challenge to human-services engineering, and the systems approach in particular, may lie in confrontation with the complexity of these major social issues—with information processing for monitoring and administrative feedback in such areas as population growth and movement, and food and resource administration; with acceptable and effective population and birth control devices and practices; with urban and suburban domiciliary construction and maintenance, averting congestion and other degrading features in life quality; with transportation and communication in centralized or decentralized urban complexes; with police and fire protection; with criminal justice and socialization; with health and mental health and public welfare.

A 1975 Gallup poll in Princeton, New Jersey, indicated that the public in both cities and the total nation considered the following problems of most current importance (listed by order of importance):

Crime
Unemployment
Transportation
Education
Poor housing
Living costs
Drugs
High taxes
Unsanitary conditions
Ineffective police
Juvenile delinquency
Lack of civic pride

Since 1949, crime went from sixth to first most important of the problems; transportation and housing remained high among the problems, with unsanitary conditions, delinquency, and high taxes relatively unchanged. New problems, of course, arise as a function of current conditions, e.g., unemployment not found as a problem in 1949. The fact that so little change in the ranking of problems occurred perhaps speaks for the status quo and the lack of confrontation of both the old and new problems.

Fixity of current systems, and resistances to change and integration, may be seen in the vested interests of commercial, professional, political and other groups militating for status quo. These will necessarily yield to episodes of crisis, perhaps only when the very survival of a system is threatened (cf., the gas shortage during the fall and winter of 1973–74). During such crises, rules and regulations become necessarily imposed of such magnitude as to limit severely the individual's freedom of operation. Stern, demanding governmental rules may indeed become the increasing consequence of multiple unyielding factors militating against change. Then may we perhaps progressively lose our democratic freedoms. Here, too, does the challenge lie for system engineers who may assist in averting such emergencies and crises through analyses that might persuade vested interest groups of the necessity for change; here, too, does the challenge lie in adapting and developing methods of analysis that will relate to the multiple facets of human-service needs in broad system terms, as well as to those who might address themselves to quick-fix problem solutions in current human-service operations.

AN AGGRESSIVE STANCE FOR THE SYSTEM ENGINEER

Over the past decade, system specialists have been increasingly entering the arena of human-service systems engineering. Technical interest groups of the Human Factors Society, for example, as a systems-oriented group have expanded their professional interests to include health, environmental design, urban problems, transportation, computers, etc. During the five-year period from 1966 through 1972, for example, the Human Factors Society technical sessions addressing human service or domestic problems increased from 28 to 73 percent of total conference content.[4] It seems increasingly apparent to such system specialists that the truly urgent contemporary problems lie in mass housing; commerce and industry; land and air transportation; fire protection; law enforcement and criminal justice; health, sanitation, mental health, and public welfare, etc.[5] Yet, the stance of those with systems expertise is often one of a timid, confused, or withdrawn nature. As a consequence, social designers or programmers simply practice covert disregard for the more total systems data, born perhaps of unknowing ignorance, if not overt avoidance, of the broader system input as an expedient economic measure.[6] Indeed, in the human-services area, human performance and acceptance and preference data are so often lacking that consumer advocacy becomes a major issue.[7]

The potential usefulness in the specialization of the systems engineer in human-service applications becomes quite obvious. The system engineer's knowledge and skills in analytic methodologies, as well as sophistication in human operator capabilities among those with human-factors expertise, provides vast potential not only for tuning up current operational systems, but for contributing to the future more total systems design of human services. Much in the future of systems engineering of human services must lie in the daring, initiative, and aggressiveness of individual system engineers who take up the challenge.

FORECASTS FOR A SYSTEMS CONTRIBUTION TO HUMAN SERVICE SYSTEMS

Modern trends tend to point to an ever increasing population growth beyond the bounds of natural feeding and health regulation and control, urban strife, resource depletion, environmental degradation, etc. These in turn point to the need for infinitely more sophisticated technologies than have been applied in the past if human societies are to operate with a modicum of efficiency, or, indeed, even to survive. Human-systems specialization may be lamented as an undirected bag of miscellania,[8] thus having no definitive future in its contribution to human services. The very survival of a human-factors system specialization may be contingent upon diversification.

Since human factors are often quite central to a system development program, a more detailed discussion of this profession may also lead to several interesting conclusions.

Beginning around the mid-1960s, aerospace human-factors activities in weapons and space-related work began to fall off. System specialists such as human engineers, industrial designers, man-in-the-loop simulation experts, and other biotechnologists in aerospace were increasingly less in demand; for their major source of occupation had previously been analysis of complex weapons and vehicles in aerospace. While some minimum effort was to continue for optimizing system performance in an aerospace environment, new efforts and areas of endeavor began to make their appearance. System designers began to turn their talents more to industrial and manufacturing systems design in optimizing the work environment, the flow of work, elimination of safety hazards, work with special tools, visual aids, and skill and training requirements in the manufacturing and quality control process. New research-and-development areas also came to the fore, with inroads being made by system engineers; man-in-the-sea studies, domestic environmental design, criminal justice and law enforcement, mass air and surface transit systems, medical and health care, urban problems, education, etc., began to make their appearance in the systems literature.[9]

Human Factors Systems Engineering, as a discipline of twenty or thirty years in the making, has consistently proved itself of implicit value in hardware cost effectiveness studies.[10] Now, however, the true worth of a human-factors contribution to domestic and human-services needs still to be meaningfully demonstrated.

The philosophy is extant throughout many areas of systems engineering expertise to work generally for the betterment of mankind. Those who join to form societies with emphasis on human needs not only have the technical skills to promote advances in outer space, but must also possess the inherent empathy and human concern to translate and adapt their skills to domestic environments for the benefit of humanity, viz., the entire world population. The nature of the multidisciplinary background required in system pursuits provides a ready-made consortium with essential skills to address all varieties of social ills; such a consortium also possesses the expertise capable of developing and implementing recommendations across the gamut of required human services; nor are the systems to be addressed implicitly limited by academic disciplines. The analytic and integrative tool—the systems approach—permits the entire complex of disciplines to relate to the more total living problems; in each instance, the system boundaries become those of the designated living environment, and the design concerns the application of social mechanisms, products, and procedures to facilitate desired outcome goals of human growth and progress.[11]

Multiple disciplines and skills will, of necessity, be required in design for the ultimate fruition of human services. Such skills and disciplines will be held together, not by a common course of study or academic degree, but by an outlook or point of view focusing on human beings as world citizens with rights to protection from the chaotic and hazardous conditions of an uncontrolled changing environment. Here, too, the methods of the systems engineer may be applied to adopt and use human resources in an efficient and effective manner; and to improve communication between social scientists and human resource technologists, to permit appropriate allocation and planning for a future society to insure availability of needed future skills for human-service developments.[12]

Current trends appear to point to an increasing involvement of systems analytic talent in consumer human-service and other domestic problems. The projection of continued and expanded performance and contributions of systems engineering would seem to indicate the following areas or arms of future system-design activities:

—Military and aerospace projects
—Conversion of major power utilization functions to alternative energy source modes for human operation
—Industrial and ergonomic applications, such as increased efficiency and humanization of work areas
—Commercial areas of concern for efficient, effective and satisfying operation of plazas, malls, retail stores, restaurants, etc.
—Information processing in analysis of requirements and parameters in management and administrative operations, and in operator interfaces with computers
—Design for population control in development of widespread acceptance and reliable usage of mechanical and chemical devices
—Participation in urban design solutions, of inner city or outlying housing developments, design for habitability, health and sanitation, food and water service, human welfare and mental health services, crime control and socialization of offenders, transportation, communication, recreation, art and creativity
—Contributions to domestic safety, emergency and preparedness provisions for natural and man-made calamities

SYSTEM PROJECTS AND METHODOLOGIES IN HUMAN-SERVICE SYSTEMS

The worthiness of future system contributions to human-service systems may perhaps be best measured by the practical input provided for the ease, reliability, and effectiveness of human-service operations. In fact, for pragmatic survival of such system professions as human factors, it may become essential to focus on desired system outcomes. Indeed, human factors and aeromedical activities had their very origin and basic success in the applied aspects of the human sciences. The human factors and other systems engineers may be well behooved to avoid purism for the sake of science, while anchoring their research to mission-oriented solutions. Research of basic perceptual problems in discrimination, etc., or research in abstract semantic differentials, etc., may often have but little payoff and remote bearing on the desired outcome of projects. Rather, definitive project developments may best take their research problems from mission-oriented human-service sytems.

A report entitled, "Project Hindsight," sponsored by the Department of Defense in 1966, traced the science and technology embodied in a number of major weapon systems. The importance of the more basic research activities in military applications was largely negated in the study. The Hindsight conclusions were, in fact, dismaying to basic-research proponents; viz., that while basic research has very little to do with development of weapon systems, mission-oriented research emphatically does. Many innovative approaches in a systems development, for example, accrue as the result of applying very fundamental principles of technology. The shopping cart, hardly removed from the principle of the wheel, made possible the self-service system in the supermarket.[13]

While sophisticated Non-Mission-Oriented Research (NMOR), or so-called basic research, has been indicated to feed into applied socio-technical applications, the overall cost variables of such research may, in fact, be indeterminate; cf., the sophisticated research on crawfish, extrasensory perception (ESP), etc. The National Science Foundation,[14] sponsoring studies at Battelle Laboratories, found that technical payoff of basic research indeed occurred, though with some more appreciable lag and higher cost than with mission-oriented research. These studies, however, failed to account for cost and resource expenditures of the vast volumes of that pure or basic research (viz., NMOR) that has achieved but minimal practical application or has *never* come to fruition.

Human-service system solutions are, of course, inordinately complex and will of necessity require voluminous supportive research. State-of-the-art and the inventory of available system components and effects may simply not be available for solution of problems in ecology, energy and the environment, sanitation and health care, mental health, transportation, housing, public welfare, etc. It is, however, from within these areas, committed to project solutions, that we must derive our research problems and developmental approaches rather than from basic research.[15]

More immediate system engineering contributions to human services may hinge upon the ability of the systems engineer to define problems and establish analytical methods within the context of broad social goals, thus establishing amenable, tractable areas for mission-oriented research, as well as providing more immediate analytical solutions.

One method which seems to have merit and widespread application potential to human-service analyses is that of "Situation Engineering." A human-factors engineer in his aerospace work is already familiar with the situation analysis approach to pilot tasks in developing control-display design and layout criteria. Mission phases, such as in a missile launch sequence, impose time and workload constraints with which time-shared human-operator task elements and console configurational components must be compatible for optimal performance. A situation in the domiciliary human-service subsystem, for example, may be a high rate of exposure of women to attack after leaving a public transit facility to enter a unit in a high-density housing project. Design for this situation should prevent extensive exposure to dark, isolated and blind areas, permitting direct entry to the unit, etc., in order to minimize likelihood of attack or rape.

Situations in the surface traffic subsystem may likewise be identified for analysis and engineering solutions. The interlocking seat belt requirement adopted by the U. S. Department of Transportation, for example, presented a number of situations for which design was not optimized, e.g., car rental, new owners, new drivers, hurried drivers, etc. The interlocks in the 1974 models, termed a tragi-comedy by many, locked in with the starter and a buzzer warning system that necessitated switch actuation with seat belt lock before ignition sequencing became possible. A retractor reel also locked in with sudden vehicle or belt acceleration, often resulting in a confused fitting and locking situation even after initial learning. Human-factors systems input was sorely needed in such situations for acceptability by even the most confusion-prone drivers. Alternative designs for effectiveness and acceptance of the imposed requirement might involve built-in seat restraints, inflatable bags at impact (a proposed feature on the 1976 and later models), etc. Trade-off criteria, as well as effective safety, must include ease, reliability, and speed of adjustment for various operational situations.

Other surface traffic engineering situations calling for systems and human engineering might include expressway night time preparatory and directional signs. Typical high-frequency situations involve anticipating, reading, and reacting to en route and destination signs in multilane, high-speed traffic flow under reduced lighting. Each anticipation and choice point requires extensive human engineering to insure optimal readability by populations with low-level acuity under adverse night-driving conditions.

Other situation analyses might concern air terminals in the air traffic subsystem. A common passenger situation is in deplaning with marginal time connections for continuation flights. Appropriate and effective signs, displays, etc., are needed at each directional choice point for readability, decision, and action while in a heavy pedestrain traffic flow pattern.

Law-enforcement situations also need to be studied more closely to optimize equipment, procedures, and training program design. For example, an aggressive situation at a neighborhood bar, etc., may be aggravated by unnecessary and bold constraints causing other semi-inebriates to lose their tempers, etc. Tranquilizing, moderating, arbitrating, etc., approaches need to be tested in designing for control of such situations in depersonalizing the arbitration and arrest procedures from the standpoint of both those who might hold the negative values for law and order, as well as the effects of harassment and stress on the enforcing police officers.

SITUATION ENGINEERING VERSUS PSYCHIATRY

Mental health problems may lend themselves to extensive situation engineering. In wartime situations, evacuation of disturbed, overwrought combat soldiers to rear echelons resulted in diagnostic reinforcement of a "mentally sick" role, which was appropriately acted out by the confused combatant. Glass[16] cites the practice of forward location and brief intensified treatment as clearly communicating to both the wearied combatant and treatment personnel that psychiatric casualties were only temporary; removal of the combatant to a rear echelon for psychiatric treatment weakened emotional ties with "buddies" (the peer group). Implications of cowardliness and weakness or failure made the sick role the only honorable way out. During the Second World War, psychiatry became effectively replaced by situation engineering. The soldier suffering fatigue and confusion, when maintained in the area of occurrence with expectancy of return, recovered in two or three days. A field medical corpsman would then lead a mute "catatonic" soldier to a shell hole. "You rest here, buddy, and you can go back with your unit in an hour or so."

The situation may merely have required temporary relief from immediate stress, good food and a bath, sleep, and a little abreaction or spilling out of the feelings he had. Above all, however, a continued expectancy of return to his unit was essential. This was the situation wherein 90 percent of the psychiatric casualties returned to their unit within three days. Evacuation for typical psychiatric treatment produced chronicity, with fewer than 10 percent returning to combat from rear echelons.

Such situation engineering has also recently been applied in civilian mental health with the use of local hospital psychiatric wards, or most notably day hospitals (see chapter 8), substituting for evacuation to remote mental hospitals as a discrete situational design or structure.

Other situational approaches in both criminal justice and mental health may be directed to the career criminal or mental patient.[17] Such career or chronic cases may be seen to have effectively rejected the common social values—the so-called criminals for reasons of reinforcing rewards outside the common social system, and the mental patient basically for reason of inability to cope with these social values. The situation-engineering approach in each case may involve behavior modification and living skills development oriented to a discharge or release target situation. The following case of a mental patient in a state hospital may be illustrative:

> Patient DW is to be discharged to an independent living situation; e.g., a tolerant boardinghouse, etc., with some possibility of employment. He evidences a number of adverse behaviors that would relate unfavorably outside in accomplishing a successful living adjustment situation, e.g., inappropriate and loud talking, easily provoked to lose his temper, etc. DW evidences a number of such problem behaviors or skill deficiencies that would be most pertinent to his discharge target. Many problem behaviors or social problems may be otherwise accepted and go uncorrected if these are not important to making an adjustment in the situation to which the patient is to be discharged. For example, a severe head-motion tic may be tolerated or accepted in the living situation (group home or boardinghouse) to which the patient will be transferred. Under the pressure of time and limited resources, then, other more cogent behaviors must be given attention, e.g., "striking people."
> Behaviors to be corrected, or skills to be imparted, must necessarily be a function of the environment in which the patient must cope after discharge. Behavior or skill deficiencies may be adjusted in one or two ways or both: orientation and training of the patient, or development of appropriate situations in the community that will accept or tolerate the patient, or for which only specific skills are required, e.g., to find work at painting for a man liking such work, or a home accepting someone who stammers, etc.

The enigma of high recidivism or readmission rates in this situation-engineering approach may be attributed to deficiencies in the correction technology, as well as in poor situational design in the discharge environment. Often behavior is seen as acceptable during the correction procedure within the institutional setting, only to be found to be inadequately controlled in the post-discharge situation. State-of-the-art in behavior modification or reinforcement therapies currently provides only partially effective outcomes, particularly since situation design is frequently wholly neglected. Moreover, superficial behavior pattern changes (cf. B. F. Skinner's behavior modification approach) may be further inadequate when failing to consider inner body processes. Recent work on biofeedback has tended to indicate that simple reinforcement of conforming behavior is not sufficient if any permanent change is to result; while even when visceral learning is involved, permanency of corrective functions without situation control is problematic.[18] Neal Miller's work at Rockefeller University in New York has demonstrated that such complex visceral functions as blood pressure can be controlled with appropriate feedback training. However, as in the case of one young woman in his study, control is lost without appropriate situational reinforcement. His young woman had learned to bring her dangerous levels of diastolic pressure well within normal range while in the laboratory. She later reverted to high-level pressures, however, when subjected to the course of a stressful life situation outside the laboratory. This is not unlike the case of those labeled mentally ill, delinquent or adult criminal, etc., who revert to previous or uncontrolled behavior modes, lacking reinforcement of corrected behaviors, or discharged into tenuous situations on the outside more conducive to eliciting "wrong" behaviors, or exposed to a situation only minimally tolerant of such—thus the high recidivism or readmissions rates.

Situation engineering, as well as other methodological approaches in the systems engineering of human-service systems, will require extensive supportive research; but the research, in order to contribute effectively to system solutions, must be mission-oriented and provide essential and improved technologies to make the systems work.

THE POLITICO-SOCIAL CLIMATE OF HUMAN SERVICES

The prevailing political, economic and social climate, and the vested interests of labor unions, professional credentialism and roles, etc., all may act as barriers to effective human-service system design. These may best be circumvented by organizational strategies developed in recent years to overcome the inertia of needy citizen groups, on the one hand, and the resistance of the reacting interest groups on the other.

Systems technology may often be applied through action programs in the democratic process, though it might seem that the principle of democracy itself is often the greatest barrier to speedy progress. Whereas some proponents of the democratic political way of life hold that democratic politics in the United States is a relatively efficient system to reinforce agreement, encourage moderation and sustain social peace among a restless immoderate people in an incredibly complex society, others deny the effectiveness of such a process:[19]

. . . As the critics of this approach are quick to say, the trouble with emphasizing democratic processes is that existing democractic practices do little to provide for the attainment of at least some unquestionably worthwhile social goals. Most importantly, who will represent the needs of all, if most political actors—individuals or groups—are self interested and narrowly concerned? . . .

Prevailing control may indeed reside in a minority, i.e., whereas 40 percent of the nation may have 60 percent of the wealth, 3 percent, constituting the upper class, control 40 percent. Interests of a group are seen as directed to the maximizing of its long-run share of values, with each group sharing its own interests. Insofar as it fails to do so, it is presumed to lack information and organization. The upper class excels in information access and organization ability in pursuing largely what must be seen as its own interests.

It is precisely in the power factions of a community where militant status quo is sought—while condemning the lazy ne'er-do-wells on welfare, or fearing crime the most since they have the most to be stolen.[20]

Those members of groups who experience or have origins in the social complex in which services have been lacking or conditions aggravated by inefficient public services,[21] often have the greatest "psychic" energy potential to be mobilized—convicted criminals who had been placed in brutal caged environments, chronic welfare recipients with low self-regard, parents of the mentally retarded, or next of kin to the mentally ill, children whose elderly parents have been institutionalized in neglectful nursing homes, those subjected to incredible traffic snarls, patients forced to wait and who are subjected to indignities or are unable to obtain meager health-care services, etc.

Political office seekers, perhaps by virtue of political gamesmanship, are frequently self-seeking, wooing those factions of the voting public who probabilistically provide the greatest voter support. They thus most often represent not the public welfare in general but major active voting blocs and pressure groups. It is, however, to these political figures that we must turn for public service programming and funding support in system design. Appropriate public policy and legislation seem essential, e.g., relating to rehabilitation rather than a dole process or incarceration, efficient people-and-commerce transit systems rather than simply road building and fueling provisions for the private auto, etc. Federal operational support requires careful systems design planning and incentive structures for local government implementation of programs.

On any given election day 50 to 60 percent of eligible voters is often considered "good" voter turnout. Thus, a tremendous slack may be evident among individuals who, if organized, could bring to bear major influence as constituents of the public servants to whom the electorate gives the power and authority to govern and regulate public and social welfare.

The extensive organization of multiple citizen pressure groups as effective voting blocs may well force the political benefactor (or purveyor of pre-election promises) to turn to a systems discipline to satisfy all factions. Organization may revolve about the following situations in which specific citizen groups are directly effected:

Slums areas
Job opportunities
Transportation, surface and air
Sanitation, solid, liquid and gaseous wastes
Health care
Mental health
Poverty, public aid, vocational opportunities, etc.
Others

Victims or members of these situations or circumstances may singly or as aggregates be committed as voting blocs and pressure groups. Such groups are normally highly concerned about their circumstances but fearful of the controlling power elite and vested interest groups.

The job of the community organizer is to maneuver the controlling factions and upper classes normally in control who may come to attack him as an enemy. Thus, he gains the confidence of those for whom he is organizing.[22] To allay fear, instill confidence, and develop mutual respect within the citizen group, is central to organizing for purposeful action. Self-interest and respect, according to community organizers, arises only out of people playing an active role in solving the crises in their lives, who are not passive, helpless recipients of public or private services. Denial of participation in necessary action to accomplish goals represents failure in the dignity of the individual and the democratic process—worse, the continued indifference and random political self-interest of those elected to public office. It is precisely in the potency of such self-interest, however, when it becomes tied in with public welfare, that the broader public system goals may be accomplished.[23]

> . . . There are enormous basic changes ahead. We cannot continue . . . in the nihilistic absurdities of our time where nothing we do makes sense. The scene around us compels us to look away quickly if we are to cling to any sanity. We are the age of pollution, progressively burying ourselves in our own waste. We announce that our water is contaminated by our own excrement, insecticides, and detergents, and then do nothing. Even a half-witted people, if

sane, would long since have done the simple and obvious—ban all detergents, develop new non-polluting insecticides, and immediately build waste-disposal units. . . . Tactics must begin within the experience of the middle class, accepting their aversion to rudeness, vulgarity and conflict. Start them easy, don't scare them off. The opposition's reactions will provide the "education" or radicalization of the middle class. It does it every time. Tactics . . . will develop in the flow of action and reaction. The chance for organization for action on pollution, inflation . . . violence, race, taxes, and other issues, is all about us. Tactics such as stock proxies and others are waiting to be hurled into the attack. . . .[24]

Organization and the winning of the middle class en masse to the causes of public welfare may make executive and legislative bodies of government increasingly aware of the need for systems technology—to become increasingly bent, in their political survival, on the need for maintaining ecological balances, viz., in the total welfare and public service system. With continuous development of citizen interest and political pressures, the following strategies may be fed back in the systems loop:

1. Providing technological leadership at the federal level to:
   City Managers
   Mayors
   Governors
   Community organizers
   Cooperating agencies
   Others
2. Soliciting support from systems technology associations and individuals
3. Developing incentive structures upon which federal funding is contingent in the accomplishment of local service goals
4. Influencing problem areas and policy of substantive professional journals toward a systems outlook
5. Sponsoring papers and sessions at professional meetings on systems implications, presented by peers in medicine, public health, housing, mass transit, mental health, welfare, etc.
6. Training community organizers for systems programs broader than the customary commitment to single unified groups

Strategies may be developed through principles of persuasion for those with self-seeking interests resistant to innovation programs:[25]

—When benefiting from status quo vested interest groups will oppose change.

—When changes threaten groups in any way they will be opposed.

—Resistence will be proportional to the pressures exerted for change.

—When key community figures or leaders accept the changes proposed, general resistances will be decreased.

—When proposed changes are seen to enhance the group's or individual's status, resistance will decrease.

—Resistance will decrease when those opposed are made part of the positive action.

—When proposed changes can be demonstrated as posing minimal threat to status, resistance will decrease.[25]

THE PRACTICAL APPLICATION OF SYSTEMS TECHNOLOGY TO HUMAN SERVICES

Federal interest and support of systems technology applied to public administration seems to ebb and flow with changing administrations; even the technologists themselves appear often to vacillate in their enthusiasms for its potential. Nonetheless, the tremendous power of a technology that can efficiently exploit the vast complexity of outer space, and develop unprecedented international weaponry must be implicitly understood to have vast potential for developing public service systems.[26] Any public service administrators or professional persons in different substantive fields might apply the technology, given a basic understanding of the approach, a high-level interest in innovation, and appropriate contacts in the power structure. Several development schemes that may be employed in persuasion of professional and other vested interest groups (given the precedence of federal leadership), may be outlined as follows:

1. Hold administrative and professional workshops, employing alternative methods of service delivery for group assessment. Weigh the personal and professional advantages in terms of yielding or gaining income, power, prestige, etc.[27]

2. Develop federally sponsored conferences in various contexts or subsystems, appealing strictly to objective technical expertise, with heavy emphasis on system implications in design and operation of public services, e.g., invite presentations or papers by civil engineers on trade-offs of road systems versus forms of mass transit to the inner city, etc.

3. Predicate federal funding of city, district and state public service programs on the use of a classical team approach. Depending upon the subsystem, of course, these might be composed of such expertise

as public administrators, union leaders, system and human-factors engineers, community organizers, appropriate engineering and agency design disciplines, etc.

THE FUTURE OF SYSTEMS TECHNOLOGY IN HUMAN SERVICES

The ever-increasing pressures of population and economic problems throughout the world, as they must truly make their impact on the United States, will likely create constantly renewed demands for improved human-service technology, both in scope and effectiveness. Indeed, order and efficiency will inevitably become primary design issues, if not simply for survival then to avoid chaos and degraded quality of life.

Human services in the present text have been addressed as a number of "subsystems" that make up a total system of service operations, from housing and transportation to law enforcement, mental health, and welfare. Each of these areas, contributing to the total system, must achieve new high levels of performance effectiveness; indeed, consensus indicates that government can no longer maintain a passive role in technology—simply screening, indexing and storing information in hope that potential users will find it is not enough. A new environment is needed for promoting the active utilization of research and technology, including explicit mission-oriented human-service programs within federal agencies financed by their line-item budgets.[28]

In the complexity of modern society, innovation is slow and unsure. As a democratic society and diverse political body, we behave predominantly as reactors rather than initiators of action. We react to energy crises; food crises; moral, urban, welfare, etc., crises, rather than carefully planning for a crisis-free (or reasonably so) environment. Technological advances must be exploited, not merely in a random fashion or as we confront crises, but intelligently planned to forestall crises and catastrophes or retrogressive life styles. Human-service needs of the future may, with careful study, be determined in population, urban, transportation, crime, welfare, health and mental health forecasts. Technological forecasts may also be brought to bear for future planning where, in the health field for example, computers may serve as diagnosticians and consulting experts for the practicing physician. By the 1990s, libraries as we know them may become obsolete and replaced by general factual and research information. Before the 1980s, urban traffic flow may be directed by computer. Briefcase computers and terminals may be available at low cost, and privately owned computers with large memories may be available by the 1980s.

Computer technology alone may vastly contribute to the effective solution of future human-service problems if properly applied,[29] and if issuing from sound and meaningful human service policies.

Effectiveness in implementing human-service policy must, of course, revolve about the validity and reality of premises upon which the policy is based. Predication of policies on false premises, such as racial supremacy, puritanical definitions of sinful behavior, mistaking retribution for rehabilitation in the criminal code, or treating the mentally ill as "sick" and punishing them for their deviant behavior (compare the eighteenth- and ninteenth-century practice of bloodletting to cure disease), etc., can only result in temporary, ineffectual human-service solutions.

Development of sound, active policies, both explicit and implicit, must be regarded as essential for the future developments of human-service technology. The part that the systems specialist may play to expedite and bring the technology to fruition must be both in analytical methodologies, in information on human and machine capabilities and limitations, and, as well, on what is provocative or reinforcing of wrong or right behaviors. The system engineer's contributions may be not only in the formulation of human needs data, once policies have been explicated, but in relating human abilities, and disabling or enabling cultural influences to the operation of extremely complex interactions in the human-service system environments of the future.

NOTES TO CHAPTER 15

1. A. Toffler, *Future Shock* (New York: Random House, 1968).

2. P. Goldmark, "New Rural Societies Experiment" (Windham Regional Development Agency, Willimantic, Connecticut, 1973).

3. E. Daddario, "The Congress and the Social Sciences," *American Journal of Orthopsychiatry*, January 1970, pp. 14–21; I. Hoos, "Systems Techniques for Managing Society: A Critique," *Public Administration Review* 33 (March 1973) :157–64.

4. "Change in Emphasis," *Human Factors Society Bulletin* 16 (February 1973) :1.

5. A. Chapanis, "Behavioral Sciences in the Nation's Service," *Human Factors Society Bulletin* 9 (December 1966) :2–3; R. McFarland, "Society President Discusses Human Factors During the Next Decade, 1970–1980," *Human Factors Society Bulletin* 13 (November 1970) :1–3.

6. A. Chapanis, "Human Factors—The Rudderless Ship," *Human Factors Society Bulletin* 12 (July 1969) :1.

7. C. Bennett, "The Adversary Role," *Human Factors Society Bulletin* 12 (May 1969) :1ff.; P. Kyropoulos, "Human Factors for the Masses," *Human Factors Society Bulletin* 14 (July 1971) :3–4; A. Zavala, "Consumerism Problems from Consumers' Viewpoints," *Human Factors Society Bulletin* 16 (May 1973) :3–5.

8. W. Miller, "A New Look at a Possible New Frontier," *Human Factors Society Bulletin* 12 (October 1969) :1–2.

9. J. Kraft, "Diversification—Key to Human Factors Survival," *Human Factors Society Bulletin* 11 (June 1968) :5–6; G. Simon, "Elementary School System Evaluation," *Human Factors Society Bulletin* 11 (August 1968) :1–4; R. Smith, "Metamorphosis: Man-in-the-Street to Man-in-the-Sea," *Human Factors Society Bulletin* 14 (September 1971) :10–12; A. Chambers and I. Seymour, "Improving the Management of Public Housing Authorities: The Role of Human Factors," Proceedings of the Seventeenth Annual Meeting of the Human Factors Society (The Human Factors Society, P. O. Box 1369, Santa Monica, California, 1973) ; J. Halldane, "Human Factors and the Environment," ibid.; D. Johnson and H. Altman, "Effects of Briefing Card Information on Passenger Behavior During Aircraft Evacuation Demonstration," ibid.; M. Rudov, "Consumer Health Education: An Analysis," ibid.; F. Zunno, "The IACP and the Police Weapons Center," ibid.

10. K. Teel, "Is Human Factors Engineering Worth the Investment?", *Human Factors Journal* 13 (February 1971) :17–21.

11. R. Gutekunst, "Manifest Destiny in Human Factors," *Human Factors Society Bulletin* 11 (February 1968) :1–2; R. Callow, "Check for Absurdity: An Overview of Where it's At," *Human Factors Society Bulletin* 11, (December 1968) :1ff.

12. G. Eckstrand et al., "Human Resources Engineering: A New Challenge," *Human Factors Journal* 9 (December 1969) :517–20.

13. J. Walsh, "Technological Innovation: New Study Sponsored by NSF," *Science* 180 (May 25, 1973) :846–47.

14. National Science Foundation, "Science, Technology and Innovation" (Washington, D.C.: Division of Science Resource Studies, 1973) .

15. "New Values for Federal Science?", *Science* 181 (August 3, 1973) :423.

16. A. Glass, "Military Psychiatry and Changing Systems of Mental Health Care," *Journal of Psychiatric Reseach* 8 (1971) :499–512.

17. D. Silber, "Controversy Concerning the Criminal Justice System and Its Implications for the Role of Mental Health Workers," *American Psychologist* 29 (April 1974) :239–44.

18. J. Jonas, *Visceral Learning. Toward a Science of Self-Control* (New York: The Viking Press, 1973) .

19. D. Ricci, "Democracy and Community Power," in *Political Power, Community and Democracy*, ed. E. Keyes and D. Ricci (Chicago: Rand McNally and Company, 1970) .

20. N. Polsby, *Community Power and Political Theory* (New Haven, Conn.: Yale University Press, 1970) ; S. Angrist, "Dimensions of Well-Being in Family Housing," *Environment and Behavior*, December 1974, pp. 495–516. In actuality, criminal acts are most to be feared among the lower classes, particularly in the slum-area high-rise domiciles.

21. In the majority of cases, some measure of service is currently operating. Often the best solution is to seek out inefficiencies, inadequacies, gaps, etc., in service, where cost-benefit improvements may be pointed up. As a matter of public record, institutions or various recourses to treatment may often aggravate, reinforce, or worsen a condition, position or situation of not only the individual

but of the social group, e.g., rising crime rate may be largely a function of first convictions and the learning of criminal behavior modes during imprisonment. Heavy doses of tranquilizers for the elderly in order to control their noisy or nuisance behavior may indeed break down control of both the individual and the discipline in the ward of a nursing home. Heavy drugs in the aged are known to be poorly excreted or metabolized, while more likely to be stored in the liver—thus resulting in retention, stupor, and persistent degrading of the elderly person's cognitive processes, his behavioral control,—thus, ward discipline.

22. S. Alinsky, *Reveille for Radicals* (New York: Random House, Inc., 1946); idem, *Rules for Radicals* (New York: Random House, Inc., 1971).

23. S. Alinsky, *Reveille for Radicals*, pp. 94f.

24. S. Alinsky, *Rules for Radicials,* pp. 191, 195.

25. M. Karlins and H. Abelson *Persuasion: How Opinions and Attitudes are Changed* (New York: Springer Publishing Company, 1970). Refer also to D. Klein, *Community Dynamics in Mental Health* (New York: John Wiley and Sons, Inc., 1968).

26. R. Boguslaw, "System Concepts in Social Systems," and C. Churchman, "A Critique of the Systems Approach to Social Organization," in *System Concepts: Lectures on Contemporary Approaches to Systems,* ed. F. Miles (New York: John Wiley and Sons, 1973), pp. 177–205.

27. A systems approach to the health-care delivery system is illustrated, focusing on the problem of existing and possible alternative modes, in M. Dade, *Modeling and Evaluation of the Health-Care Delivery System* (Santa Monica, Calif.: Rand Corporation, 1973).

28. "Giving Spin to Technology Spin-offs," *Science News* 105 (May 18, 1974): 318.

29. Parsons and Williams, "Forecasts of 1968–2000 of Computer Developments," *Nyropsgade* 47 (1602 Copenhagen V, Denmark, 1968).

Glossary

AFTERCARE: medical service often neglected in patient follow-up treatment subsequent to inpatient status. In mental health, sometimes termed "sustaining care," referring to an effort to sustain the patient in the community.

CARDIAC FIBRILLATION: runaway heart action losing rhythm and regularity. Death results in a few minutes unless excited to recovery.

CODED DEPARTMENT: established operational agency with specific and restricted charter of service.

COHORT(ING): specified groups or their formation as aggregates of patients or service recipients with a common basis in space and time. For example, a group discharged from, or entering, the same institution during the same period may be designated as cohorts. Such grouping provides for controlled evaluation studies. Also applied to socially cohesive groups.

COMMUNITY CARE-GIVING: system of health, mental health, and other services delivered within the immediate community obviating removal to remote institutional settings.

COMMUNITY MENTAL-HEALTH ACT: federal law enacted in 1963 (Public Law 88–164) to provide for staffing and construction seed money in developing local clinics and comprehensive services.

COMPONENT: an operational entity within system operational boundaries. For example, a fire alarm is a component of a fire-fighting and control system.

COMPUTER JUSTIFICATION: establishing value of computer application, in trade-off with manual, etc., processing based on cost and efficiency criteria.

CONSTRAINTS: restrictions and limitations within which system developments must occur. Funding limitations would, for example, be a constraint in developing a human-service program.

COUPLING MODES: means by which interactive functional elements are accomplished in the design of human services. The method of collecting fees from the public in a mass transit service, through a guard at a gate, coin-actuated turnstile, etc., is a coupling-mode problem.

CRITICAL INCIDENT: technique developed by John Flannagan to identify significant areas and conditions of human error in aircraft control. Method may be adapted to human-service systems analysis to identify areas of failure.

285

DESIGN SPECIFICATIONS: documented detailed requirements for construction or development of a functional system of services oriented to the accomplishment of definitive human-service goals.

EXTENDED CARE FACILITIES: group-care homes in the community organized and operated for the retention of geriatric, mentally ill, retarded or otherwise socially or physically debilitated persons. Often criticized for custodial rather than rehabilitative orientation.

EXTRUSION: practice of expelling or removing troubled or troublesome persons from the community.

FAMILY HEALTH CARE: community health services designed to meet the total needs of families, ranging from pediatric to geriatric care, medical, dental, psychiatric, etc.

FUNCTIONS ANALYSIS: classical developmental approach, translating system requirements into generalized operational descriptions. A requirement to enter a human-service system may, for example, be translated into a reception function.

GENERALIST SERIES: a set of human-service professional specifications formulated by the Illinois Department of Mental Health upon which training requirements and job function certification are predicated.

GOAL-DIRECTED APPROACHES: system formulation and design based on specified outcomes to be accomplished. For example, a design might be predicated on a goal to develop productive employment situations for 25 percent of a mentally ill population within a one-year time frame.

GOALS: desired status, ends or outcomes a system is designed to accomplish. Goals are often erroneously oriented to the functions themselves rather than to the effective outcomes. Thus, to develop recreational facilities is a function rather than an outcome oriented to the improved use of leisure time. The latter, as an accomplishment, may be measured as an outcome parameter such as increased frequency of participation by an elderly population, etc.

HEALTH MAINTENANCE ORGANIZATION (HMO): federally sponsored system of health care composed of a specified user population, a funding source or arrangement, and a group of cooperating practitioners.

HUMAN FACTORS: putting the user into the design of things used. This includes design of suitable services for the users as well as appropriate tasks for human operator performance in providing the services.

INPATIENT: a resident or bedded patient in hospital or mental hospital.

INPUT STATE: condition (s) of debility, deficiency, social maladjustment, distress or other needs which the system is to serve in accomplishing correction, rehabilitation, alleviation or improvement criteria.

INTEGRATION: bringing all operational components together, adjusting and coordinating these to accomplish total system output requirements within required limits of accuracy.

INTERIOR DESIGN ANALYSIS: building design based on systematic behavioral or user criteria, rather than intuitively developed notions of what is required.

INTERVENTION: breaking a cycle or trend otherwise ultimately leading to condition (s) of debility, deficiency, social maladjustment or distress.

MANAGEMENT: awareness and adeptness in current management science state-of-the-art in personnel administration, computer application and justification, project monitoring cost effectiveness, integration and evaluation.

MASS TRANSIT: transportation system of common carriers to transport large volumes of materials and personnel in short, intermediate and long-range travel. Modes may include land, sea, air, orbital, and subterranean.

MEANS APPROACH: distinct from a requirements approach which specifies what needs to be accomplished rather than how. Welfare systems, for example, have generally precisely laid out approaches for service to the poor.

MENTAL: designation often given in legalese and law enforcement for a "mentally ill" person, or one evidencing deviant behaviors not otherwise to be interpreted as delinquent or criminal.

MISSION: goal-oriented developmental activities.

MISSION ANALYSIS: detailed sequential development leading to the solution of system operational requirements.

MISSION-ORIENTED RESEARCH: research projects taking substantive problem content from system-operational requirements and leading to system solutions—distinct from basic research which is not oriented to immediate or developmental applications.

MODEL: a set of pictorial, graphic or symbolic relationships permitting study of real-world situations. These are exemplified by wind tunnels, mockups, mathematical formulas, etc.

NETWORK ANALYSIS: a mathematical model used as engineering analytical methods in the study of flow processes. Used in the study of critical paths to complete projects or applied to patient flow paths, etc.

NONRESIDENTIAL LODGE: psychiatric social club in the community designed for extensive cohort supportive participation of chronic mental patients.

HUMAN-SERVICE NONSYSTEM: disjointed character of present service agencies each going their separate ways.

OPERATIONS RESEARCH: analytic approach developed during World War II to optimize operational effectiveness.

OUTPATIENT: medical or psychiatric case treated through office or home visit. Contrasted with "inpatient" or residential case.

OUTPUT STATE: desired status or situation which system purports to accomplish, i.e., cure, rehabilitation, reduced transit time, etc.

PARTIAL HOSPITALIZATION: day treatment of psychiatric patient wherein patient returns to own abode at night. Also known as day care, not to be confused with child-care centers.

POLICY: implicit or explicit official position upon which developmental goals and practices are based in such areas as national defense, public welfare, public health, mental health, public transit, etc.

POLICY ISSUES: differing views concerning basic approaches to human-service problems. For example, "get tough" custodial approaches to juvenile delinquency may contrast with freedom and occupational training approaches.

PREVENTION: intervention approaches to physical and mental health, juvenile delinquency, etc., designed to reduce the incidents.

PRIMARY MEDICAL CARE: basic health services in the treatment of illness, correction of visual defects, nutrition, dentistry, etc.

PROCESS ORIENTATION: primary concern and preoccupation with conventional practices rather than desired outcomes for which other than conventional approaches may be more effective.

PSYCHIATRY: medical practice in treating the "psyche" or aberrated behavior through medication, psychoanalytic audits, etc.

QUANTITATIVE AND QUALITATIVE PERSONNEL REQUIREMENTS INFORMATION (QQPRI): analytic procedure developed for the U.S. Air Force to determine skills and training necessary for personnel in effective system operation.

QUICK-FIX: approaches to solution of current operational problems for immediate alleviation in human-service systems.

REACTION TIME: time requirements or constraints for a system to operate effectively, e.g., in fire fighting, crime checks, etc.

RECLAMATION: restored use of land, lakes, and other resources made unusable due to neglect, contamination and pollution.

RECYCLING: processing liquid, solid, and gaseous waste products to become useful resources once again.

REHABILITATION: to restore to useful or socially acceptable behavior or functions.

RELIABILITY: consistency of data. Replication of procedures or repeated interrogations for different points of view evidencing some measure of the same results (cf. VALIDITY) .

REQUEST FOR PROPOSAL (R FOR P): document outlining system performance requirements and specifications and directed toward potential bidders.

SANITATION: public health function concerned with disease control and prevention.

SIMULATION MODELS: pictorial or symbolic representations of a system with algorithms or real-world relationships programmed to depict system operations.

SOCIAL COHESIVENESS: theoretical affinity and supportive interest among members forming a cohort or group coming from mental hospitals, etc.

SOCIAL INDICATORS: measures obtained from a setting or milieu in which the problems are defined, e.g., crimes-of-violence frequency.

SOURCE OF REFERRAL: agency, group or person sending, directing or advising another in obtaining an available human service.

SUBSYSTEM: group of components operating to a common purpose and oriented in its operation to an even broader range of objectives or goals, e.g., juvenile recreation center in a delinquency prevention system.

SYSTEM BOUNDARIES: essential definitions circumscribing the processes and dynamics involved in accomplishing outcome objectives.

SYSTEM DESIGN OBJECTIVES: specified outcome measures to which the accomplishment of integrated operational components are oriented, e.g., percent increase of productive labor force from mentally ill and psychopathic populations, etc.

SYSTEM OPERATIONAL REQUIREMENTS (SOR): policy-based, broad-ranging operational and quantifiable outcome goals about which system design specifications develop, e.g., to reduce criterion measures of substantive poverty among the American Indians within a decade.

TASK ANALYSIS: analytic methods formulated by Robert Miller in the early 1950s to determine specific man-machine relationships involving required skills and training requirement. Adaptable to human-service systems.

THERAPY: treatment, correction or curing processes.

TIME LINE: analytic base upon which to program time-constrained mission activities.

TRAINING FUNCTIONS ANALYSIS: determination of required task elements and the unavailable skills necessary for training in mission performance and the most efficient methods for imparting these, i.e., on the job training, classroom instruction, etc.

TRAINING PROGRAM DESIGN: comprehensive plan for developing training functions in completing the necessary human-operator skills inventory for mission completion.

VALIDITY OF SERVICE: a measure of the meaningful character of functions that are directed toward accomplishment of system outcome goals, e.g., psychoanalysis may fail to upgrade the living states of lower socioeconomic groups, or further highway construction may fail to alleviate inner city congestion, etc.

VESTED INTERESTS/CREDENTIALISM: professional or economic status which may be threatened by innovative approaches to human service solutions, e.g., subprofessional medical treatment may usurp the M.D.'s business, effective mass transit systems threaten the automobile manufacturer with reduced sales, etc.

Bibliography

Ackoff, R. L., and Sasieni, M. W. *Fundamentals of Operations Research.* New York: John Wiley & Sons, 1970.

Alinsky, S. *Rules for Radicals.* New York: Random House, Inc., 1971.

Allison, David. *The R&D Game.* Boston: The M.I.T. Press, 1969.

Ayres, Robert U. *Technological Forecasting and Long-Range Planning.* New York: McGraw-Hill Book Company, 1969.

Barker, Roger, and Schoggen, Philip. *Qualities of Community Life.* San Francisco, Calif.: Jossey-Bass Publishers, 1973.

Bauer, Raymond A., ed. *Social Indicators.* Boston: The M.I.T. Press, 1966.

Bell, Daniel, ed. *Toward the Year 2000: Work in Progess.* Boston: Beacon Press, 1969.

Bell, G.; Randall, E.; and Roeder, J. *Urban Environments and Human Behavior.* Stroudsburg, Pa.: Dowden Hutchinson & Ross, Inc., 1973.

Berry, Brian. *City Classification Handbook.* New York: John Wiley & Sons, 1972.

Black, Guy. *Application of Systems Analysis to Government Operations.* New York: Praeger Publishers, 1968.

Blesser, William B. *A Systems Approach to Biomedicine.* New York: McGraw-Hill Book Company, 1969.

Branch, Melville. *Planning Urban Environments.* Stroudsburg, Pa.: Dowden, Hutchinson & Ross, Inc., 1974.

Bright, James R., ed. *Technological Forecasting for Industry and Government.* Englewood Cliffs, N.J.: Prentice-Hall, Inc., 1968.

Buckley, Walter. *Sociology and Modern Systems Theory.* Englewood Cliffs, N.J.: Prentice-Hall, Inc., 1967.

Catanese, Anthony. *Scientific Methods of Urban Analysis.* Urbana, Ill.: University of Illinois Press, 1972.

Chacko, George K. *Applied Statistics in Decision Making.* New York: American Elsevier, 1971.

Chapanis, Alphonse. *Ethnic Variables in Human Factors Engineering.* Baltimore, Md.: The Johns Hopkins University Press, 1975.

Chapanis, Alphonse. *Research Techniques in Human Engineering.* Baltimore, Md.: The Johns Hopkins University Press, 1959.

Chestnut, Harold. *Systems Engineering Methods.* New York: John Wiley & Sons, 1967.

Churchman, C.; West, R.; Ackoff, R.; and Arnoff, E. *Introduction to Operations Research.* New York: John Wiley & Sons, 1957.

Cleland, D. *Systems Analysis and Project Management.* New York: McGraw-Hill Book Company, 1968.

Cox, D. R., and Miller, H. D. *The Theory of Stochastic Processes.* New York: John Wiley & Sons, 1966.

Dade, M. *Modeling and Evaluation of the Health Care Delivery System.* Santa Monica, Calif.: Rand Corporation, 1973.

Dixon, J. R. *Design Engineering Inventiveness, Analysis, and Decision Making.* New York: McGraw-Hill Book Company, 1966.

Dober, Richard. *Environmental Design.* New York: Van Nostrand Reinhold Company, 1969.

Dole, S. H. et al. *Establishment of a Long-Range Planning Capability.* Santa Monica, Calif.: Rand Corporation RM6151 NASA, NAS 2-5459, September 1969.

Drucker, P. *The Practice of Management.* New York: Harper and Row, Publishers, 1954.

Drummond, M. *System Evaluation and Measurement Techniques.* Englewood Cliffs, N.J.: Prentice-Hall, Inc., 1971.

Easton, David. *A Systems Analysis of Political Life.* New York: John Wiley & Sons, 1965.

Eckman, D. P., ed. *Systems Research and Design.* New York: John Wiley & Sons, 1961.

Eddison, R. T.; Pennycuick, K.; and Rivett, B. M. P. *Operational Research in Management.* New York: John Wiley & Sons, 1962.

Ellis, David O., and Ludwig, Fred J. *Systems Philosophy.* Englewood Cliffs, N.J.: Prentice-Hall, Inc., 1962.

English, J. Morley, ed. *Cost Effectiveness.* New York: John Wiley & Sons, 1968.

Enke, Stephen, ed. *Defense Management.* Englewood Cliffs, N.J.: Prentice-Hall, Inc., 1967.

Euston, Andrew. "Community Development and Environmental Design." *HUD Challenge,* August 1975, pp. 6–8.

Ewald, William. *Environment for Man: The Next Fifty Years.* Bloomington, Ind.: Indiana University Press, 1970.

Faludi, Andreas. *Planning Theory.* New York: Pergamon Press, 1973.

Feldman, M. L.; Gonzalez, L. A.; and Nadel, A. B. *Application of Aerospace Technologies to Urban Community Problems.* Santa Monica, Calif.: General Electric Company NASA Purchase Order No. 5177, July 1965.

Fisher, Gene. *Cost Considerations in Systems Analysis.* New York: American Elsevier, 1970.

Flagle, C. D.; Huggins, W. H.; and Roy, R. H. *Operations Research and Systems Engineering.* Baltimore, Md.: The John Hopkins University Press, 1960.

Forrester, J. W. *Urban Dynamics*. Boston: The M.I.T. Press, 1969.

Gagne, Robert. *Psychological Principles in System Development*. New York: Holt, Rinehart and Winston, 1962.

Gallion, Arthur, and Eisner, Simon. *The Urban Pattern: City Planning and Design*. New York: Van Nostrand Reinhold Company, 1963.

Godschalk, David. "Participation, Planning and Exchange in Old and New Communities: A Collaborative paradigm." Chapel Hill, N.C.: University of North Carolina Center for Urban and Regional Studies, 1971.

Goldman, Thomas A., ed. *Cost-Effectiveness Analysis: New Approaches in Decision Making*. New York: Praeger Publishers, 1967.

Goode, Harry H., and Machol, Robert E. *System Engineering*. New York: McGraw-Hill Book Company, 1957.

Gordon, Geoffrey, *System Simulation*. Englewood Cliffs, N.J.: Prentice-Hall, Inc., 1969.

Hajek, Victor G. *Project Engineering*. New York: McGraw-Hill Book Company, 1965.

Hall, Arthur D. *A Methodology for Systems Engineering*. New York: Van Nostrand Publishers, 1962.

Helmer, Olaf. *Social Technology*. New York: Basic Books, 1966.

Heywood, Philip. *Planning and Human-Needs*. New York: Praeger Publishers, 1973.

Iberall, Arthur. *Toward a General Science of Viable Systems*. New York: McGraw-Hill Book Company, 1972.

Johnson, R. A.; Kast, F. E.; and Rosenzweig, J. E. *The Theory and Management of Systems*. New York: McGraw-Hill Book Company, 1967.

Kalmus, H. *Regulation and Control in Living Systems*. New York: John Wiley & Sons, 1966.

Karlins, M., and Abelson, H. *Persuasion: How Opinions and Attitudes are Changed*. New York: Springer Publishing Company, 1970.

Kast, Fremont E., and Rosenzweig, James E. *Science, Technology, and Management*. New York: McGraw-Hill Book Company, 1963.

Katz, Daniel, and Kahn, Robert. *The Social Psychology of Organizations*. New York: John Wiley & Sons, 1966.

Kelleher, Grace J., ed. *The Challenge to Systems Analysis: Public Policy and Social Change*. New York: John Wiley & Sons, 1970.

Klir, George J., ed. *Trends in General Systems Theory*. New York: John Wiley & Sons, 1972.

Lott, Richard W. *Basic Systems Analysis*. New York: Harper and Row, Publishers, 1971.

Machol, Robert E., ed. *System Engineering Handbook*. New York: McGraw-Hill Book Company, 1965.

McKean, Roland N. *Efficiency in Government Through Systems Analysis.* New York: John Wiley & Sons, 1958.

Mesarovic, Mihajlo. *Views on General Systems Theory.* New York: John Wiley & Sons, 1964.

Mesarovic, M. *Systems Approach and the City.* New York: American Elsevier, 1972.

Michael, Donald. *The Unprepared Society: Planning for a Precarious Future.* New York: Basic Books, 1972.

Miles, F. ed. *System Concepts: Lectures on Contemporary Approaches to Systems.* New York: John Wiley & Sons, 1973.

Miller, David W., and Starr, Martin K. *Executive Decisions and Operations Research.* Englewood Cliffs, N.J.: Prentice-Hall, Inc., 1969.

Mitford, J. *Kind and Unusual Punishment.* New York: Alfred A. Knopf, Inc., 1973.

Morse, Philip M., and Bacon, Laura W., ed. *Operations Research for Public Systems.* Boston: The M.I.T. Press, 1967.

Morton, J. A. *Organizing for Innovation: A Systems Approach to Technical Management.* New York: McGraw-Hill Book Company, 1971.

National Commission on Urban Problems. *Building the American City: Report of the National Commission on Urban Problems.* New York: Frederick A. Praeger, Publishers, 1971.

Naylor, T. H. et al. *Computer Simulation Techniques.* New York: John Wiley & Sons, 1966.

Pelz, Donald D., and Andrews, Frank M. *Scientists in Organizations.* New York: John Wiley & Sons, 1966.

Peterson, E. L. *Statistical Analysis and Optimization of Systems.* New York: John Wiley & Sons, 1961.

Quade, E. S., and Boucher, W. I. *Systems Analysis and Policy Planning.* New York: American Elsevier, 1968.

Ramo, Simon. "New Dimensions of Systems Engineering." In *Science and Technology in the World of the Future,* edited by A. B. Bronwell. New York: John Wiley & Sons, 1970.

Revelle, Roger, and Landsberg, Hans, eds. *America's Changing Environment.* Boston: Houghton Mifflin Company, 1970.

Ricci, D. "Democracy and Community Power." In *Political Power, Community and Democracy,* edited by E. Keynes and D. Ricci. Chicago: Rand McNally and Company, 1970.

Rothenberg, Jerome. *The Measurement of Social Welfare.* Englewood Cliffs, N.J.: Prentice-Hall, Inc., 1961.

Sandler, G. H. *System Reliability and Engineering.* Englewood Cliffs, N.J.: Prentice-Hall, Inc., 1963.

Seiler, Karl. *Introduction to Systems Cost Effectiveness.* New York: John Wiley & Sons, 1969.

Shinners, Stanley M. *Techniques of System Engineering.* New York: McGraw-Hill Book Company, 1967.

Simon, Herbert. *The New Science of Management Decision.* New York: Harper and Row, Publishers, 1960.

Tocher, K. D. *The Art of Simulation.* New York: Van Nostrand, 1963.

Truxal, John G. *Introductory Systems Engineering.* New York: McGraw-Hill Book Company, 1972.

Tufte, Edward. *The Quantitative Analysis of Social Problems.* Reading, Mass.: Addison-Wesley Publishing Company, 1970.

Tuwiner, Sidney, ed. *Environmental Science Technology Information Resources.* Park Ridge, N.J.: Noyes Data Corporation, 1973.

Vayda, Andrew, ed. *Environment and Cultural Behavior.* New York: The American Museum of Natural History, The Natural History Press, 1969.

Wagner, Harvey M. *Principles of Operations Research with Applications to Managerial Decisions.* Englewood Cliffs, N.J.: Prentice-Hall, Inc., 1969.

Walton, Thomas F. *Technical Data Requirements for Systems Engineering and Support.* Englewood Cliffs, N.J.: Prentice-Hall, Inc., 1965.

Weisman, H. *Overcoming Mismanagement in the Human Service Profession.* San Francisco, Calif.: Jossey-Bass., Publishers, 1973.

Williams, J. D. *The Compleat Strategyst.* New York: McGraw-Hill Book Company, 1954.

Wilson, Warren E. *Concepts of Engineering System Design.* New York: McGraw-Hill Book Company, 1965.

Wynkoop, Sally. *Directories of Government Agencies.* Rochester, N.Y.: Rochester Libraries Unlimited, 1969.

Zadeh, L. A., and Polak, E. *System Theory.* New York: McGraw-Hill Book Company, 1969.

Index